Contents

Conference Board of the Mathematical Sciences
REGIONAL CONFERENCE SERIES IN MATHEMATICS

supported by the
National Science Foundation

Number 53

CLOSED GEODESICS ON RIEMANNIAN MANIFOLDS

by
WILHELM KLINGENBERG

Published for the
Conference Board of the Mathematical Sciences
by the
American Mathematical Society
Providence, Rhode Island

Expository Lectures
from the CBMS Regional Conference
held at the University of Florida
August 2–6, 1982

1980 *Mathematics Subject Classifications.* Primary 53C22, 58B20, 58D15, 58E10.

Library of Congress Cataloging in Publication Data

Klingenberg, Wilhelm, 1924–

Closed geodesics on Riemannian manifolds.

(Regional conference series in mathematics; no. 53)

"Expository lectures from the CBMS regional conference held at the University of Florida, August 2–6, 1982" – T. p. verso.

Bibliography: p.

1. Riemannian manifolds. 2. Curves on surfaces. I. Conference Board of the Mathematical Sciences. II. Title. III. Series.

QA1.R33 no. 53 [QA649] 510s [514′.74] 83-5979

ISBN 0-8218-0703-X

ISSN 0160-7642

Printed in the United States of America

Introduction

We consider compact Riemannian manifolds M. Every tangent vector X to M determines a geodesic $c(t) = c_X(t)$, $t \in \mathbf{R}$, by the condition $\dot{c}(0) = X$. If $X \neq 0$ and if there exists an $\omega > 0$ such that $\dot{c}(\omega) = \dot{c}(0)$, then $c \mid [0, \omega]$ is called a closed geodesic.

Let $\tau \equiv \tau_M : TM \to M$ be the tangent bundle of M. On TM we have the geodesic flow $\phi_t : TM \to TM$. Here, $\phi_t X$ is given by $\dot{c}_X(t)$.

The geodesics flow is a Hamiltonian flow. More precisely, consider the cotangent bundle $\tau_M^* : T^*M \to M$. On T^*M we have a canonical symplectic structure $\alpha^* = -d\theta$ where θ is the canonical 1-form given by $-\sum v^i du^i$ in the local coordinates (u^i, v^i) of T^*M. Moreover, from the Riemannian metric g^* on T^*M we have the function

$$E^* : T^*M \to \mathbf{R}, \quad X^* \to \tfrac{1}{2} g^*(X^*, X^*).$$

The cogeodesic flow is the flow of the Hamiltonian system (T^*M, α^*, E^*). In the local coordinates of T^*M, the Hamiltonian equations read

$$\frac{du^i}{dt} = E^*_{v^i} = \sum_l g^{il}(u) v^l; \qquad \frac{dv^i}{dt} = -E^*_{u^i} = -\frac{1}{2} \sum_{k,l} \frac{\partial g^{kl}(u)}{\partial u^i} v^k v^l.$$

Using the canonical isomorphism

$$\begin{array}{ccc} TM & \xrightarrow{L_g} & T^*M \\ \downarrow{\scriptstyle \tau_M} & & \downarrow{\scriptstyle \tau_M^*} \\ M & \xrightarrow{\mathrm{id}} & M \end{array} ; \qquad \begin{array}{ccc} (u^i, \dot{u}^k) & \longmapsto & (u^i, \Sigma_k g_{ik}(u) \dot{u}^k) \\ \downarrow & & \downarrow \\ (u^i) & \longmapsto & (u^i) \end{array}$$

we can transport the cogeodesic system (T^*M, α^*, E^*) into the geodesic system (TM, α, E). The Hamiltonian equations of this system read, in local coordinates,

$$\frac{du^i}{dt} = E_{v^i} = \dot{u}^i, \quad \frac{d\dot{u}^i}{dt} = -E_{u^i} = -\sum_{j,k} \Gamma^i_{jk}(u) \dot{u}^j \dot{u}^k.$$

That is to say, the flow lines $\phi_t X$ project under τ_M into the geodesics on M.

We thus see that the closed geodesics are in $1:1$ correspondence with periodic flow lines on $TM - M$. Or, in the language of mechanics: Geodesics correspond to motions of a point in the geodesic system and, in particular, closed geodesics correspond to periodic motions.

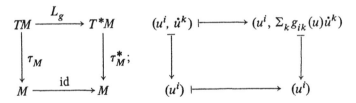

Clearly, when considering motions we can restrict ourselves to those of unit speed. That is, we can restrict the geodesic flow to the unit tangent space T_1M of M.

From the two aspects of a closed geodesic—a closed curve which is a geodesic and a geodesic (flow line) which is periodic—we elaborate the first one. We will consider a space formed by all closed curves in which the closed geodesics are characterized as the critical points of a functional. This idea goes back to Morse. In our exposition we will give a refined version of Morse's approach which has several advantages over the old one—in particular, it possesses a canonical $\mathbf{O}(2)$-action.

Chapter 1. The Hilbert manifold of H^1-curves

We begin here the construction of a proper Hilbert manifold, associated canonically to a finite-dimensional Riemannian manifold M. The Hilbert manifold $H^1(I, M)$ in question is formed by the maps $c: I \to M$ of class H^1, $I = [0, 1]$; cf. 1.1. Of greater importance than the space of all H^1-maps $c: I \to M$ are certain submanifolds of finite codimension which we will introduce in Chapter 2.

Here we describe in full detail the natural atlas for $H^1(I, M)$. The charts $(\exp_c^{-1}, \mathfrak{U}(c))$, $c \in H^1(I, M)$, are defined with the help of the inverse of the exponential map, restricted to 'short' vector fields along c; cf. 1.10, 1.12.

Next we describe the two canonical bundles α^0 and α^1 over $H^1(I, M)$. The fibre $(\alpha^0)^{-1}(c)$ over c consists of the H^0-vector fields along c, while $(\alpha^1)^{-1}(c)$, consisting of the H^1-vector fields along c, is identified with the tangent space $T_c H^1(I, M)$ at c; cf. 1.13. This is then used to show that there exists a canonical Riemannian metric on $H^1(I, M)$ (cf. 1.19), and the energy integral $E(c)$ is differentiable; cf. 1.20. the only critical points of E are the constant maps.

It is well known that the set $C^0(I, \mathbf{R}^n)$ of continuous curves $c: I = [0, 1] \to \mathbf{R}^n$ in Euclidean n-space $\mathbf{R}^n$ is a Banach space. $C^0(I, \mathbf{R}^n)$ can be viewed as the completion of the space $C'^{\infty}(I, \mathbf{R}^n)$ of piecewise differentiable curves with respect to the maximum norm

$$\|c\|_\infty = \sup_{t \in I} |c(t)|\,.$$

The associated distance is

$$d_\infty(c, c') = \sup_{t \in I} |c(t) - c'(t)|\,.$$

For an element $c \in C^0(I, \mathbf{R}^n)$, in general, there exists neither the length $L(c)$ nor the energy integral $E(c)$. Therefore, in differential geometry one considers a different norm on $C'^{\infty}(I, \mathbf{R}^n)$, i.e., the norm $\|c\|_1$ derived from the scalar product

$$\langle c, c'\rangle_1 = \langle c, c'\rangle_0 + \langle \partial c, \partial c'\rangle_0.$$

Here, $\langle e, e'\rangle_0 = \int_I e(t) \cdot e'(t)\, dt$ and $\partial c(t) = \dot{c}(t)$. The completion of $C'^{\infty}(I, \mathbf{R}^n)$ with respect to the norm $\|c\|_1$ is denoted by $H^1(I, \mathbf{R}^n)$. According to a classical result of Lebesgue, an H^1-curve $c: I \to \mathbf{R}^n$ can be described as an absolutely continuous curve for which $\partial c(t) = \dot{c}(t)$ exists for almost all t and $\partial c(t)$ is square integrable, i.e., $\langle \partial c, \partial c \rangle_0 < \infty$.

We recall that $c: I \to \mathbf{R}^n$ is called absolutely continuous if, for every $\varepsilon > 0$, there exists a $\delta > 0$ such that $0 \leqslant t_0 < \cdots < t_{2k+1} \leqslant 1$ and $\Sigma_{i=0}^k | t_{2i+1} - t_{2i} | < \delta$ imply $\Sigma_i | c(t_{2i+1}) - c(t_{2i}) | < \varepsilon$.

Note that, in particular, $H^1(I, \mathbf{R}^n)$ is a subset of $C^0(I, \mathbf{R}^n)$. Actually, as one can easily see, the inclusion is continuous; cf. 1.5. The functional $E(c) = \langle \partial c, \partial c \rangle_0 / 2$ on $C'^\infty(I, \mathbf{R}^n)$ possesses an extension to the Hilbert space $H^1(I, \mathbf{R}^n)$ and there it defines an $\mathbf{R}$-valued C^∞-function. Its differential is given by $DE(c).\eta = \langle c, \eta \rangle_0$.

We now extend these constructions to the case of curves on a n-dimensional Riemannian manifold M. We assume M to be complete, $I = [0, 1]$.

1.1 Definition. (i) Denote by $C'^\infty(I, M)$ the set of piecewise differentiable curves $c: I \to M$.

(ii) By $C^0(I, M)$ denote the space of continuous curves $c: I \to M$, endowed with the metric

$$d_\infty(c, c') = \sup_{t \in I} d(c(t), c'(t)).$$

Here, d is the metric derived from the Riemannian metric on M.

(iii) A curve $c: I \to M$ is said to be of class H^1 if, for a chart (u, M') of M and $I' = c^{-1}(M')$, the mapping $t \in I' \mapsto u \circ c(t) \in \mathbf{R}^n$ is in $H^1(I', \mathbf{R}^n)$. By $H^1(I, M)$ we denote the set of H^1-maps $c: I \to M$.

Note. In 1.1(iii) the particular choice of the chart does not play a role. Indeed, if $\phi: U \subset \mathbf{R}^n \to U' \subset \mathbf{R}^n$ is a diffeomorphism and $u: J \to U$ is of class H^1, then $\phi \circ u: J \to U'$ is also of class H^1.

We have the following canonical inclusions: $C'^\infty(I, M) \hookrightarrow H^1(I, M) \hookrightarrow C^0(I, M)$.

1.2 Proposition. *$C'^\infty(I, M)$ is a dense subspace of the complete metric space $\{C^0(I, M), d_\infty\}$.*

Proof. A curve $c \in C^0(I, M)$ can be covered by finitely many charts. Thus, the proposition is reduced to the well-known fact that $C'^\infty(I', \mathbf{R}^n)$ is dense in $C^0(I', \mathbf{R}^n)$, I' an interval in I.

On $H^1(I, M)$ we can consider the energy integral E.

1.3 Proposition. *For $c \in H^1(I, M)$,*

$$E(c) = \frac{1}{2} \int_1 \langle \dot{c}(t), \dot{c}(t) \rangle \, dt$$

is well defined.

Proof. Let $u \circ c: I' \to \mathbf{R}^n$ be a local representation of $c | I'$ with respect to some chart (u, M'). Then we can define $2E(c | I')$ to be the integral over I' of the function $g(u \circ c(t))((u \circ c)^{\cdot}(t), (u \circ c)^{\cdot}(t))$. Note that $u \circ c \in H^1(I', \mathbf{R}^n)$ possesses a square integrable derivative. In the same way as for curves of class C^∞, one shows that this is independent of the particular choice of the chart.

As a preparation for the introduction of the structure of a differentiable manifold on $H^1(I, M)$, modelled on a Hilbert space, we first consider the infinitesimal approximation of $C'^{\infty}(I, M)$ at $c \in C'^{\infty}(I, M)$ given by the vector space of piecewise differentiable vector fields along c. This vector space can be viewed as the 'tangent space' $T_cC'^{\infty}(I, M)$ of the 'manifold' $C'^{\infty}(I, M)$ at the point c. Our main interest lies in its completion with respect to an H^1-norm. To get a precise formulation, we take $T_cC'^{\infty}(I, M)$ to be the space of sections in the induced bundle $c^*\tau$.

1.4 DEFINITION. Let $c \in C'^{\infty}(I, M)$.

(i) Define $c^*\tau$ to be the induced bundle over I:

$$\begin{array}{ccc} c^*TM & \xrightarrow{\tau^*c} & TM \\ \downarrow{\scriptstyle c^*\tau} & & \downarrow{\scriptstyle \tau} \\ I & \xrightarrow{c} & M \end{array}$$

For each closed subinterval $I_j \subset I$ with $c \mid I_j$ differentiable, the restriction of $c^*\tau$ to I_j is the (differentiable) induced bundle. The fibre $c^*\tau^{-1}(t)$ over t is also denoted by $T_{c,t}$.

(ii) By $C'^{\infty}(c^*TM)$ we denote the vector space of piecewise differentiable sections of $c^*\tau$. We also write $T_cC'^{\infty}(I, M)$ instead and call this the tangent space to $C'^{\infty}(I, M)$ at c.

(iii) Using the scalar product $\langle\, , \rangle$ on the fibres $T_{c,t}$ of $c^*\tau$, stemming from the Riemannian metric on the corresponding $T_{c(t)}M$, we define, for ξ, η in $C'^{\infty}(c^*TM)$:

(a) $$\|\xi\|_{\infty} = \sup_{t \in I} |\xi(t)|,$$

(b) $$\langle \xi, \eta \rangle_0 = \int_I \langle \xi(t), \eta(t) \rangle \, dt,$$

(c) $$\langle \xi, \eta \rangle_1 = \langle \xi, \eta \rangle_0 + \langle \nabla\xi, \nabla\eta \rangle_0.$$

The norm derived from the scalar product $\langle\, , \rangle_r$ is denoted by $\|\ \|_r$, $r = 0, 1$.

(iv) The completion of $C'^{\infty}(c^*TM)$ with respect to the norms $\|\ \|_{\infty}$ and $\|\ \|_r$ is denoted by $C^0(c^*TM)$ and $H^r(c^*TM)$, $r = 0, 1$, respectively.

REMARK. A piecewise differentiable section of $c^*\tau$ is a continuous map $\xi : I \to c^*TM$ with $c^*\tau \circ \xi = \mathrm{id}$ and such that there exists a subdivision of I into closed intervals J with $c \mid J$ differentiable and $\xi \mid J$ differentiable. For $\xi \mid J$ we have the covariant derivative $\nabla\xi \mid J$ with respect to the induced connection on $c^*\tau \mid J$. Alternatively, ξ, or rather $\tau^*c \circ \xi$, can be viewed as a piecewise differentiable vector field along c.

Note that $C^0(c^*TM)$ is a Banach space, whereas $H^r(c^*TM)$, $r = 0, 1$, are Hilbert spaces.

1.5 PROPOSITION. *The inclusions* $H^1(c^*TM) \hookrightarrow C^0(c^*TM) \hookrightarrow H^0(c^*TM)$ *are continuous. More precisely*:

(i) *if* $\xi \in C^0(c^*TM)$, *then* $\|\xi\|_0 \leqslant \|\xi\|_\infty$; *and*

(ii) *if* $\xi \in H^1(c^*TM)$, *then* $\|\xi\|_\infty \leqslant \sqrt{2}\,\|\xi\|_1$.

PROOF.

(i) $$\|\xi\|_0^2 = \int_I \langle \xi(t), \xi(t)\rangle\, dt \leqslant \int_I \max_t |\xi(t)|^2\, dt = \|\xi\|_1^2.$$

(ii) Choose $t_1 \in I$ with $\|\xi\|_\infty = |\xi(t_1)|$. Then

$$\|\xi\|_\infty^2 = |\xi(t)|^2 + \int_t^{t_1} \frac{d}{ds}|\xi(s)|^2\, ds \leqslant |\xi(t)|^2 + 2\int_I |\xi(s)|\,|\nabla\xi(s)|\, ds$$

$$\leqslant \|\xi\|_0^2 + \|\xi\|_0^2 + \|\nabla\xi\|_0^2 \leqslant 2\|\xi\|_1^2. \quad \square$$

Note. Besides vector bundles of the type $c^*\tau : c^*TM \to I$, we will also be led to consider associated vector bundles like product bundles, bundles of multilinear mappings, etc.

In each case, such a bundle $\pi : E \to I$ carries a Riemannian metric and a Riemannian connection. This allows us to form, just as in 1.4 for the bundle $c^*\tau$, the completions $C^0(E)$, $H^0(E)$, $H^1(E)$ of the bundle $C'^\infty(E)$ of piecewise differentiable sections in π. 1.5 also holds in this case. The proof is exactly the same.

1.6 DEFINITION. Let $\pi : E \to I$; $\phi : F \to I$ be vector bundles with Riemannian metric and Riemannian connection. Let $\mathcal{O} \subset E$ be open such that $\mathcal{O}_t = \pi^{-1}(t) \neq \varnothing$, all $t \in I$.

(i) Denote by $H^1(\mathcal{O})$ the set of $\xi \in H^1(E)$ with $\xi(t) \in \mathcal{O}_t$, all $t \in I$.

(ii) Let $f : \mathcal{O} \to F$ be a differentiable fibre mapping. That is, for each $t \in I$, $f|\mathcal{O}_t$ is differentiable with image in $F_t = \phi^{-1}(t)$. Then define $\tilde{f} : H^1(\mathcal{O}) \to H^1(F)$; $(\xi(t)) \mapsto (f \circ \xi(t))$. Our next goal is to show that $H^1(\mathcal{O}) \subset H^1(E)$ is open and $\tilde{f}$ is differentiable. We begin with the

1.7 PROPOSITION. *With the notions of* 1.6, $\tilde{f}$ *is a continuous mapping defined on an open subset of* $H^1(E)$.

PROOF. Denote by $C^0(\mathcal{O})$ the set of continuous sections ξ of π with $\xi(t) \in \mathcal{O}_t$. Then there exists a $\rho > 0$ such that $\xi \in C^0(\mathcal{O})$, $\eta \in C^0(E)$, $\|\xi - \eta\|_\infty < \rho$ implies $\eta(t) \in \mathcal{O}_t$, all $t \in I$. Hence, $C^0(\mathcal{O})$ is open in $C^0(E)$. $H^1(\mathcal{O})$ is the counterimage of $C^0(\mathcal{O})$ under the continuous inclusion $H^1(E) \to C^0(E)$. Therefore, $H^1(\mathcal{O})$ is open. To see that f is continuous we note that with $\|\eta - \xi\|_1$ small, $\|\eta - \xi\|_\infty$ and $\|\nabla\eta - \nabla\xi\|_0$ also become small. Moreover,

$$\|\tilde{f}(\eta) - \tilde{f}(\xi)\|_0 \leqslant \|\tilde{f}(\eta) - \tilde{f}(\xi)\|_\infty.$$

Now let $\dot\eta(t) = \dot\eta(t)_h + \dot\eta(t)_v$ be the decomposition into the horizontal and the vertical part, respectively. $\dot\eta(t)_h$ has the local representation $(t, \eta(t), \partial t, -\Gamma_t(\partial t, \eta(t)))$, and $\dot\eta(t)_v$ is canonically identified with $\nabla\eta(t)$. Therefore, if

$$T_{\eta(t)} f = T_{\eta(t),1} f + T_{\eta(t),2} f$$

denotes the decomposition into the restrictions upon the horizontal and vertical subspaces, respectively, we have

$$\nabla(f\circ\eta)(t)-\nabla(f\circ\xi)(t)=T_{\eta(t),2}f.\nabla\eta(t)-T_{\eta(t),2}f.\nabla\xi(t)$$
$$=T_{\eta(t),2}f.\left(\nabla\eta(t)-\nabla\xi(t)\right)+\left(T_{\eta(t),2}f-T_{\xi(t),2}f\right).\nabla\xi(t).$$

Hence, with $\|\xi-\eta\|_1\to 0$, $\|\nabla\tilde{f}(\eta)-\nabla\tilde{f}(\xi)\|_0\to 0$ also. □

1.8 PROPOSITION. *Let $\pi: E\to I$; $\phi: F\to I$ be vector bundles as in* 1.7. *Let $L(\pi;\phi): L(E;F)\to I$ be the associated bundle of linear mappings with the induced Riemannian metric and Riemannian connection. Then the canonical inclusions $H^1(L(E;F))\hookrightarrow L(H^0(E);H^0(F))$ and $H^1(L(E;F))\hookrightarrow L(H^1(E);H^1(F))$, given by*

$$A=(A(t))\mapsto\left\{\tilde{A}:=(\xi(t))\mapsto(A(t).\xi(t))\right\},$$

are continuous. More precisely,

$$(*)\qquad \|\tilde{A}(\xi)\|_0^2\leqslant\|A\|_\infty^2\|\xi\|_0^2\leqslant 2\|A\|_1^2\|\xi\|_0^2$$

and

$$(**)\qquad \|\tilde{A}(\xi)\|_1^2\leqslant 8\|A\|_1^2\|\xi\|_1^2.$$

PROOF. We use 1.5. Then $(*)$ is obvious. To prove $(**)$ we note that

$$\nabla A(\xi)(t)=\nabla A(t).\xi(t)+A(t).\nabla\xi(t).$$

Hence,

$$\|\nabla\tilde{A}(\xi)\|_0^2=\|\nabla A.\xi+A.\nabla\xi\|_0^2\leqslant 2\|\nabla A.\xi\|_0^2+2\|A.\nabla\xi\|_0^2$$
$$\leqslant 4|\nabla A|_0^2\|\xi\|_1^2+4\|A\|_1^2\|\xi\|_1^2\leqslant 8\|A\|_1^2\|\xi\|_1^2.\quad\square$$

REMARK. A similar result holds for the canonical inclusions

$$H^1(L(E_1,E_2,\ldots,E_k;F))\hookrightarrow L(H^0(E_1),H^1(E_2),\ldots,H^1(E_k);H^0(F))$$

and

$$H^1(L(E_1,E_2,\ldots,E_k;F))\hookrightarrow L(H^1(E_1),H^1(E_2),\ldots,H^1(E_k);H^1(F)).$$

1.9 LEMMA. *Let $f:\mathcal{E}\to F$ be a fibre map; cf.* 1.6. *Then $\tilde{f}: H^1(\mathcal{E})\to H^1(F)$ is once continuously differentiable with the tangential given by $T_2f^\sim$.*

Note. The differentiability of arbitrarily high order is proved along the same lines starting with the differentiable fibre map $T_2f:\mathcal{E}\to L(E;F)$. See Klingenberg [**3**] and Flaschel und Klingenberg [**1**] for details. 1.9 is a slight generalization of a result contained in Palais [**1**] and Eliasson [**1**].

PROOF. From 1.7 we know that $\tilde{f}$ is continuous. The Taylor formula gives

$$f(\eta(t))-f(\xi(t))-T_{\xi(t),2}f.(\eta(t)-\xi(t))=r(\xi(t),\eta(t)).(\eta(t)-\xi(t)).$$

Here,

$$r(\xi(t),\eta(t))=\int_1 T_{\xi(t)+s(\eta(t)-\xi(t)),2}f\,ds-T_{\xi(t),2}f$$

is a fibre map of some $\mathcal{O}' \times \mathcal{O}' \subset \mathcal{O} \times \mathcal{O}$, $\mathcal{O}'$ open, into the bundle $L(\pi; \phi): L(E; F) \to I$.

Consider the associated continous mapping $r: H^1(\mathcal{O}' \times \mathcal{O}') \to H^1(L(E; F))$. Then, with $T_2 f: H^1(\mathcal{O}') \to H^1(L(E; F))$,

$$\|\tilde{f}(\eta) - \tilde{f}(\xi) - T_2\tilde{f}(\xi).(\eta - \xi)\|_1 = \|\tilde{r}(\xi, \eta).(\eta - \xi)\|_1$$
$$\leqslant \text{const}\|\tilde{r}(\xi, \eta)\|_1 \|\eta - \xi\|_1.$$

Since $\tilde{r}(\xi, \xi) = 0$ we get $\|\tilde{r}(\xi, \eta)\|_1 \to 0$ with $\|\xi - \eta\|_1 \to 0$, i.e., $T_2 f^{\sim} \in H^1(L(H^1(E); H^1(F)))$ has the properties of a differential, $T\tilde{f} = T_2 f^{\sim}$. □

There exists an open neighborhood $\mathcal{O}$ of the 0-section of $\tau: TM \to M$ with the property that $\exp|(\mathcal{O}_p = T_pM \cap \mathcal{O})$ is a diffeomorphism onto its image = open neighborhood of $p \in M$. For M compact, we can choose $\mathcal{O}$ to be the ε-ball bundle of τ, some small $\varepsilon > 0$, i.e., $\mathcal{O}_p = B_\varepsilon(0_p)$, all $p \in M$.

1.10 Definition. Let $\mathcal{O}$ be as above. For $c \in C'^{\infty}(I, M)$ denote by $\mathcal{O}_c$ the subset of c^*TM, formed by the $\mathcal{O}_{c,t} = \mathcal{O}_c \cap T_{c,t}$, which corresponds under τ^*c to $T_{c(t)}M \cap \mathcal{O}$.

Define

$$(\dagger) \qquad \exp_c: H^1(\mathcal{O}_c) \to H^1(I, M)$$

by $(\xi(t)) \mapsto (\exp_{c(t)}\tau^*c\xi(t))$ and denote the image by $\mathcal{U}(c)$.

1.11 Proposition. $(\dagger)$ *is bijective. Let $c, d \in C'^{\infty}(I, M)$.*

$$\exp_d^{-1} \circ \exp_c : \exp_c^{-1}(\mathcal{U}(c) \cap \mathcal{U}(d)) \to \exp_d^{-1}(\mathcal{U}(d) \cap \mathcal{U}(c))$$

is a diffeomorphism between open sets in the Hilbert spaces $H^1(c^*TM)$ and $H^1(d^*TM)$.

Proof. $\mathcal{U}(c)$ consists precisely of those $e \in H^1(I, M)$ with $e(t) \in \exp_{c(t)}(\mathcal{O}_{c(t)})$. This shows that $(\dagger)$ is a bijection.

For each $t \in I$ we form

$$\mathcal{O}_{c,d,t} = \mathcal{O}_{c,t} \cap (\exp \circ \tau^*c)^{-1} \circ (\exp \circ \tau^*d)\mathcal{O}_{d,t}$$

and put $\bigcup_{0 \leqslant t \leqslant 1} \mathcal{O}_{c,d,t} = \mathcal{O}_{c,d}$ if $\mathcal{O}_{c,d,t} \neq \varnothing$ for all $t \in I$. Otherwise put $\mathcal{O}_{c,d} = \varnothing$. $\mathcal{O}_{c,d}$ is an open subset of $\mathcal{O}_c$, and

$$H^1(\mathcal{O}_{c,d}) = \exp_c^{-1}(\mathcal{U}(c) \cap \mathcal{U}(d)).$$

The map

$$f_{d,c}: (\exp \circ \tau^*d)^{-1} \circ (\exp \circ \tau^*c) : \mathcal{O}_{c,d} \to d^*TM$$

is a fibre map, $\exp_d^{-1} \circ \exp_c = f_{d,c}^{\sim}$. Hence, 1.9 applies. □

1.12 Theorem. *The set $H^1(I, M)$ of the H^1-mapping $c: I \to M$ is a Hilbert manifold; its differentiable structure is given by the natural atlas $\{\exp_c^{-1}, \mathcal{U}(c); c \in C'^{\infty}(I, M)\}$.*

Proof. The charts are modelled on a proper separable Hilbert space with typical representative $H^1(c^*TM) \cong H^1(I, \mathbf{R}^n)$.

The family $\mathfrak{U}(c)$, $c \in C'^{\infty}(I, M)$, is an open covering of $H^1(I, M)$ since $C'^{\infty}(I, M)$ is dense in $H^1(I, M) \subset C^0(I, M)$; see 1.2. 1.11 shows that the natural atlas is differentiable.

To see that $H^1(I, M)$ has a countable base it suffices to show that the natural atlas has a countable subatlas. Actually, it suffices to show that, for a sequence $\{M^k\}$ of relatively compact open subsets M^k of M with $\bigcup_k M^k = M$, $M^k \subset M^{k+1}$, there exists a countable subatlas of the natural atlas covering $H^1(I, M^k)$. To see this we show that for fixed k and any integer $l > 0$ the set $H^1(I, M^k)^l = \{c \in H^1(I, M^k);\ E(c) < l\}$ can be covered by a finite subset of the natural atlas. To prove this we choose $\iota = \iota(M^k) > 0$ to be the injectivity radius on M^k. Let $m = m(\iota, l)$ be an integer satisfying $18l < m\iota^2$. Then $e \in H^1(I, M^k)^l$ implies, for $t \in [(j-1)]/m, j/k]$,

$$d\left(e\left(\frac{j-1}{m}\right), e(t)\right)^2 \leqslant \left(\int_{(j-1)/m}^{j/m} |\dot{e}(t)|\, dt\right)^2 \leqslant \frac{1}{m}\int_I \langle \dot{e}(t), \dot{e}(t)\rangle\, dt \frac{2l}{m} \leqslant \frac{\iota^2}{9}.$$

Hence, $e \mid [(j-1)/m, j/m]$ lies entirely in an $\iota/3$-ball.

There exists a finite set P of points on M^k such that the $\iota/3$-balls around these points will cover M^k. Given $e \in H^1(I, M^k)^l$, we can find a sequence $\{p_1, \ldots, p_m\}$ in P such that $e(j/m) \in B_{\iota/3}(p_j)$. For each of these finitely many sequences we choose a $c \in C'^{\infty}(I, M)$ such that $c \mid [(j-1)/m, j/m]$ is the minimizing geodesic from p_{j-1} to p_j. Then $e \in H^1(I, M^k)$ implies $e \in \mathfrak{U}(c)$, for one of these c's.

REMARK. Given a differentiable mapping $f\colon M \to N$ from a manifold M into a manifold N, we obtain a mapping

$$H^1(I, f)\colon H^1(I, M) \to H^1(I, N); \qquad (c(t)) \mapsto (f \circ c(t)).$$

This mapping is differentiable. And if $g\colon N \to L$ is another differentiable mapping we get $H^1(I, g \circ f) = H^1(I, g) \circ H^1(I, f)$. Thus, $H^1(I,\ )$ constitutes a covariant functor from the category {finite-dimensional manifold and differentiable mappings} into the category {Hilbert manifolds and differentiable mappings}.

Associated with the manifolds $H^1(I, M)$ there are in a natural way two vector bundles $\alpha^r\colon H^r(H^1(I, M)^*TM) \to H^1(I, M)$, $r = 0, 1$. The total space is the union of the spaces $H^r(c^*TM)$, $c \in H^1(I, M)$. α^1 is canonically isomorphic to the tangent bundle $\tau_{H^1(I, M)}$ of $H^1(I, M)$. To make this precise we define for $\xi \in \mathcal{O}$, $\mathcal{O}$ an open neighborhood of M in TM as above,

$$T_{\xi,1}\exp = T_\xi \exp \circ \left(T\tau \mid T_{\xi_h}TM\right)^{-1}\colon T_{\tau\xi}M \to T_{\exp\xi}M,$$

$$T_{\xi,2}\exp = T_\xi \exp \circ \left(K \mid T_{\xi_v}TM\right)^{-1}\colon T_{\tau\xi}M \to T_{\exp\xi}M.$$

Here, $T\tau \mid T_{\xi_h}TM\colon T_{\xi_h}TM \to T_{\tau\xi}M$ and $K \mid T_{\xi_v}TM\colon T_{\xi_v}TM \to T_{\tau\xi}M$ are the canonical isomophisms. Under our assumptions, $T_{\xi,1}\exp$ and $T_{\xi,2}\exp$ are linear isomorphisms.

Note. In the above definitions, we have written exp instead of $\exp_p = \exp \mid T_pM$, where $p = \tau\xi$. There can be no misunderstanding as to where the base point of the vector is. A similar simplification will also be used further down.

Define, for $c \in C^{\prime\infty}(I, M)$ and $r = 0, 1$,

$$\phi_{r,c}^{-1} : H^1(\mathcal{O}_c) \times H^r(c^*TM) \to (\alpha^r)^{-1}(c)$$

by

$$(\xi(t), \eta_c(t)) \to \left((T_{\tau^* c\xi(t),2} \exp).\tau^* c\eta_c(t)\right).$$

Here, the right-hand side is viewed as an H^r-mapping $I \to TM$, which under τ goes into the base H^1-curve ($\exp \circ \tau^* c\xi(t)$) belonging to $\mathfrak{U}(c)$.

1.13 LEMMA. *The family* $\{(\phi_{r,c}, \exp_c^{-1}, \mathfrak{U}(c)); c \in C^{\prime\infty}(I, M)\}$ *consitutes a bundle atlas for a bundle* α^r *over* $H^1(I, M)$ *associated with the natural atlas of* $H^1(I, M)$. *The typical fibre of the bundle is the separable Hilbert space* $H^r(I, \mathbf{R}^n)$. *The bundle* α^1 *is canonically isomorphic to the tangent bundle* $\tau_{H^1(I, M)}$.

PROOF. First consider the case $r = 1$. Then we see that, for c, d in $C^{\prime\infty}(I, M)$,

$$\phi_{1,d} \circ \phi_{1,c}^{-1} : H^1(\mathcal{O}_{c,d}) \times H^1(c^*TM) \to H^1(\mathcal{O}_{d,c}) \times H^1(d^*TM)$$

is of the form

$$\left(\exp_d^{-1} \circ \exp_c, T(\exp_d^{-1} \circ \exp_c)\right) \equiv (\tilde{f}_{c,d}, T\tilde{f}_{c,d}),$$

with $f_{c,d}$ as in the proof of 1.11. This shows that the above atlas is precisely the tangent atlas associated with the natural atlas of $H^1(I, M)$.

When $r = 0$, we observe that the composition maps $\phi_{0,d} \circ \phi_{0,c}^{-1}$ are again of the form $(\tilde{f}_{d,c}, T\tilde{f}_{d,c})$, and the composition mapping

$$H^1(\mathcal{O}_{c,d}) \overset{T\tilde{f}_{d,c}}{\to} H^1(L(c^*TM; d^*TM)) \to L(H^0(c^*TM); H^0(d^*TM))$$

is differentiable; see 1.8. $\square$

Note. The previous result shows that we obtain an intrinsic description of the tangent space $T_e H^1(I, M)$ of $H^1(I, M)$ at an arbitrary element $e \in H^1(I, M)$ by considering the vector space of H^1-maps $\eta : I \to TM$ satisfying $\tau \circ \eta = e$. That is, η is an H^1-vector field along the H^1-curve e.

Before we prove that we also have a natural scalar product on $T_e H^1(I, M)$, we show that natural charts exist for every $e \in H^1(I, M)$. This will follow from the next lemma. To formulate our result we put $\tilde{\mathcal{O}} = \{\eta \in TH^1(I, M); \eta(t) \in \mathcal{O}\}$, with $\mathcal{O} \subset TM$ as before.

1.14 LEMMA. *The mapping*

$$\tilde{F} = \tau_{H^1(I, M)} \times \widetilde{\exp} : \tilde{\mathcal{O}} \subset TH^1(I, M) \to H^1(I, M) \times H^1(I, M);$$
$$\eta(t) \mapsto (\tau \circ \eta(t), \exp \eta(t))$$

is differentiable. It maps a sufficiently small open neighborhood $\tilde{\mathcal{O}}' \subset \tilde{\mathcal{O}}$ *of the zero section of* $\tau_{H^1(I, M)}$ *onto an open neighborhood of the diagonal of* $H^1(I, M) \times H^1(I, M)$.

PROOF. Using the local representation of $T\mathfrak{U}(c)$ (see 1.13), we obtain for $\widetilde{\exp}$ the representation

$$(\xi(t), \eta_c(t)) \mapsto \left((\exp \circ \tau^*c)^{-1} \circ \exp \circ T_{\tau^*c\xi(t),2} \exp.\tau^*c\eta_c(t)\right).$$

This is a differentiable fibre map from $\mathcal{O}_c \times c^*TM$ into c^*TM. Now apply 1.9.

$\tilde{F}$ maps the zero section of $\tau_{H^1(I,M)}$ bijectively onto the diagonal of $H^1(I, M) \times H^1(I, M)$. It only remains to show that $T\tilde{F}$ at $O_c \in T_cH^1(I, M)$ is a bijection. But this follows by looking at the local representation: It carries $(\xi(t), 0)$ into $(\xi(t), \xi(t))$ and $(0, \eta(t))$ into $(0, \eta(t))$. □

1.15 COROLLARY. *For every $e \in H^1(I, M)$ there exists a natural chart $(\exp_e^{-1}, \mathfrak{U}(e))$ with*

$$\exp_e \equiv \widetilde{\exp} \mid \tilde{\mathcal{O}}' \cap T_eH^1(I, M) : \tilde{\mathcal{O}}' \cap T_eH^1(I, M) \to \mathfrak{U}(e). \quad \square$$

1.16 PROPOSITION. *The mapping*

$$\partial : H^1(I, M) \to H^0(H^1(I, M)^*TM); \quad (e(t)) \mapsto (\partial e(t) \equiv \dot{e}(t))$$

is a differentiable section in the bundle α^0. For $e \in \mathfrak{U}(c)$, $\xi = \exp_c^{-1}e$, the representation of ∂e in the bundle chart over $\mathfrak{U}(c)$, is given by

$$\partial_c\xi(t) = \nabla\xi(t) + \theta_c\xi(t)$$

with

$$\theta_c\xi(t) = \tau^*c^{-1}(T_{\tau^*c\xi(t),2}\exp)^{-1} \circ (T_{\tau^*c\xi(t),1}\exp) \circ \tau^*c.\partial t.$$

PROOF. We have

$$\begin{aligned}\partial e(t) &= T_{\tau^*c\xi(t)}\exp.(\tau^*c\xi(t)_h + \tau^*c\xi(t)_v) \\ &= (T_{\tau^*c\xi(t),1}\exp) \circ \tau^*c.\partial t + (T_{\tau^*c\xi(t),2}\exp) \circ \tau^*c.\nabla\xi(t).\end{aligned}$$

This gives the expressions for $\partial_c\xi(t)$ and $\theta_c\xi(t)$. In particular, $\theta_c : \mathcal{O}_c \to c^*TM$ is a fibre mapping. Hence, the mapping which associates to $\xi = \exp_c^{-1}e \in H^1(\mathcal{O}_c)$ the principal part, $\nabla\xi + \tilde{\theta}_c\xi \in H^0(c^*TM)$, of the representation of ∂e is differentiable. □

1.17 THEOREM. *The bundle α^0 of H^0-vector fields along H^1-curves on M has a Riemannian metric which is characterized by the property that on $(\alpha^0)^{-1}(c) = H^0(c^*TM)$, $c \in C'^{\infty}(I, M)$, it is given by $\langle\ ,\ \rangle_0$. We therefore denote this metric by $\langle\ ,\ \rangle_0$ in general.*

PROOF. With $\mathcal{O}$ as above define $G : \mathcal{O} \subset TM \to L(TM, TM)$ by

$$\langle G(\xi), \ \rangle_{\tau(\xi)} = \langle T_{\xi,2}\exp, T_{\xi,2}\exp\rangle_{\exp\xi}.$$

From the properties of $\mathcal{O}$ it follows that $G(\xi)$ is a positive selfadjoint operator of class C^∞.

$$G_c : (\tau^*c)^{-1} \circ G \circ (\tau^*c) : \mathcal{O}_c \to L(c^*TM, c^*TM)$$

is a fibre map. Thus, the composite mapping

$$H^1(\mathcal{E}_c) \xrightarrow{\tilde{G}_c} H^1(L(c^*TM; c^*TM)) \to L(H^0(c^*TM); H^0(c^*TM))$$

is a positive selfadjoint operator of class C^∞; it therefore defines a Riemannian metric on the representation $H^1(\mathcal{E}_c) \times H^0(c^*TM)$ of $(\alpha^0)^{-1}\mathcal{U}(c)$. Clearly, this metric does not depend on the particular representation. □

As a preparation for defining a Riemannian metric also on $\alpha^1 = \tau_{H^1(I,M)}$, we prove an analogue of 1.10.

1.18 PROPOSITION. *The Levi-Cività covariant derivation ∇ on M determines a Riemannian covariant derivation ∇_{α^0} on the bundle α^0. In particular, if η is a vector field on $H^1(I, M)$, the mapping $\eta \mapsto \nabla_{\alpha^0}(\partial).\eta$ is a section in α^0. Since, for $\alpha^1(\eta) \in C'^\infty(I, M)$, it coincides with the H^0-vector field $\nabla\eta(t)$ along $c(t)$, we also write $\nabla\eta$ instead of $\nabla_{\alpha^0}(\partial).\eta$.*

PROOF. Denote by $\Gamma(\xi)$, $\xi \in \mathcal{E}$, the Christoffel symbol of the Levi-Cività connection in the normal coordinates based at $\tau\xi \in M$. Let $c \in C'^\infty(I, M)$. Then we get a bundle mapping

$$\Gamma_c : \mathcal{E}_c \to L(c^*TM, c^*TM; c^*TM)$$

by

$$\Gamma_c = (\tau^*c)^{-1} \circ (\Gamma \circ \tau^*c)(\tau^*c \times \tau^*c).$$

The associated mapping (cf. 1.9)

$$\begin{aligned}\tilde{\Gamma}_c : H^1(\mathcal{E}_c) &\to H^1(L(c^*TM, c^*TM; c^*TM)) \\ &\to L(H^1(c^*TM), H^0(c^*TM); H^0(c^*TM))\end{aligned}$$

represents the desired Riemannian connection ∇_{α^0}.

To compute $\nabla_{\alpha^0}(\partial).\eta$ we consider the bundle trivializations of α^0 and α^1 over $(\exp_c^{-1}, \mathcal{U}(c))$. According to 1.13, $\eta \in (\alpha^1)^{-1}(e)$ is represented by $(\xi(t), \eta_c(t))$, with $\xi = \exp_c^{-1} e$. 1.16 gives the representation $\partial_c\xi(t)$ of ∂e. Thus, the principal part of the representation of $\nabla_{\alpha^0}(\partial).\eta$ reads

$$D_2(\partial_c\xi(t)).\eta_c(t) + \Gamma_c(t)(\eta_c(t), \partial_c\xi(t)).$$

Substituting 1.16 for $\partial_c\xi(t)$ we get

$$\nabla\eta_c(t) + D_2(\theta_c\xi(t)).\eta_c(t) + \Gamma_c(t)(\eta_c(t), \partial_c\xi(t)).$$

Thus, if $e = c$, i.e., if $\xi = 0$, this simplifies to $\nabla\eta_c(t)$. □

1.19 THEOREM. *The Hilbert manifold $H^1(I, M)$ has a Riemannian metric; for each element $c \in C'^\infty(I, M) \subset H^1(I, M)$, the scalar product on $T_cH^1(I, M) \cong H^1(c^*TM)$ coincides with the product $\langle\ ,\ \rangle_1$. We denote this Riemannian metric on $H^1(I, M)$ by $\langle\ ,\ \rangle_1$.*

PROOF. Define the Riemannian scalar product on $T_e H^1(I, M)$ by $(\xi, \eta) \mapsto \langle \xi, \eta \rangle_0 + \langle \nabla\xi, \nabla\eta \rangle_0$. That this is a differentiable section in $L_s^2(\alpha^1)$ follows from 1.16 and 1.18. □

We conclude this section with the

1.20 THEOREM. *The energy integral* $E: H^1(I, M) \to \mathbf{R}$ *is differentiable, with* $DE(c).\eta = \langle \partial c, \nabla\eta \rangle_0$.

PROOF. The differentiability of E follows from 1.16 and 1.17. To determine its differential we need only recall from the proof of 1.18 that the local representation of $\nabla\eta$, for $\eta \in T_c H^1(I, M)$, yields $\nabla_{\alpha^0}\partial c.\eta = \nabla\eta$. That is, $D(\frac{1}{2}\langle \partial c, \partial c \rangle_0).\eta = \langle \partial c, \nabla\eta \rangle_0$. □

1.21 COROLLARY. *The only critical points of* E *on* $H^1(I, M)$ *are the constant maps.*

PROOF. Clearly, $c =$ const, i.e., $\partial c = 0$ implies $DE(c) = 0$. Conversely, $DE(c).\eta = \langle \partial c, \nabla\eta \rangle_0 = 0$, for all $\nabla\eta \in H^0(c^*TM)$, implies $\partial c = 0$. □

Note. This reflects the fact that $H^1(I, M)$ possesses a canonical retraction onto $M =$ the space of constant maps,

$$H^1(I, M) \times I \to H^1(I, M); \qquad (\{c(t)\}, s) \mapsto \{c(t(1-s))\}.$$

Only for certain submanifolds of $H^1(I, M)$ do there exist nontrivial critical points of E. Regarding this, see the next chapter.

Chapter 2. The loop space and the space of closed curves

In this section we consider certain geometrically important submanifolds of the Hilbert manifold $H^1(I, M)$, introduced in Chapter 1. The most important ones are the loop space $\Omega_{pq}M = \Omega M$ of H^1-curves going from a fixed point $p \in M$ to a fixed point $q \in M$ and the space of closed H^1-curves. Putting $[0, 1]/\{0, 1\} = S$, we also write $H^1(S, M)$ or ΛM, for this space.

The critical points of $E \mid \Omega_{pq}M$ are the geodesics from p to q and the critical points of $E \mid \Lambda M$ are, besides the constant maps, the closed geodesics; cf. 2.3. Ω and Λ are complete with respect to the distance derived from the Riemannian metric; cf. 2.7. It can also be shown that these spaces are geodesically complete. Of great importance is the fact that the gradient vector field grad E on ΩM and on ΛM satisfies condition (C) of Palais and Smale—for ΛM one must assume here that M is compact; cf. 2.9. This condition is a substitute for the failure of a proper Hilbert manifold to be locally compact.

We can now construct critical points with the help of the minimax method on appropriate families, called ϕ-families, of subsets of ΩM or ΛM; cf. 2.17 and 2.18. As a first application we show that E assumes its infimum on every connected component of ΩM or ΛM. This implies, e.g., that in every class of freely homotopic closed curves, not homotopic to a constant curve, there exists a shortest closed curve which is a closed geodesic (2.19). In 2.20 we show that on a simply connected compact manifold there also exists at least one closed geodesic.

We continue with the concepts and notations introduced in Chapter 1. We start with the following general observation:

2.1 PROPOSITION. *The mapping* $P\colon H^1(I, M) \to M \times M$; $c \mapsto (c(0), c(1))$ *is a submersion. As a consequence, if* $N \subset M \times M$ *is a submanifold of codimension* k, $H^1_N M = P^{-1}(N)$ *is a submanifold of* $H^1(I, M)$ *of codimension* k.

PROOF. The differentiability of P is clear from the local representation. The tangential T_cP at c is given by

$$\eta \in T_cH^1(I, M) \cong H^1(c^*\tau) \mapsto (\eta(0), \eta(1)) \in T_{c(0)}M \times T_{c(1)}M,$$

which shows that P is a submersion. □

2.2 DEFINITION. The submanifold $P^{-1}(N)$ of $H^1(I, M)$ is denoted by $H^1_N M$. By $C^0_N M$ we denote its completion in $C^0(I, M)$.

For the special case $N = \{p, q\}$ we get the H^1-loop space $\Omega_{pq}M$ or ΩM. For $N = \Delta =$ diagonal of $M \times M$ we get the space ΛM of (parameterized) closed curves.

REMARKS. 1. The tangent space $T_c\Omega M$ of $\Omega M = \Omega_{pq}M$ at c can be identified with the H^1-vector fields $\xi(t)$ along $c(t)$ satisfying $\xi(0) = \xi(1) = 0$. The tangent space $T_c\Lambda M$ consists of the periodic H^1-vector fields $\xi(t)$ along $c(t)$, i.e., $\xi(0) = \xi(1)$.

2. Among the various other submanifolds $H^1_N M = P^{-1}(N)$ of $H^1(I, M)$ we only mention the case $N = M_0 \times M_1 \subset M \times M$, where M_0, M_1 are submanifolds of M. In this case, $H^1_N M$ is formed by the H^1-curves c with $c(0) \in M_0$, $c(1) \in M_1$. $T_c H^1_N M$ consists of the $\xi(t)$ along $c(t)$ with $\xi(0) \in T_{c(0)}M_0$, $\xi(1) \in T_{c(1)}M_1$.

We can now characterize the critical points of E in $H^1_{M_0 \times M_1}M$ and ΛM.

2.3 LEMMA. (i) *The critical points c of $E \mid H^1_{M_0\times M_1}M$ are the geodesics with $c(0) \in M_0$, $\dot c(0) \in T^\perp_{c(0)}M_0$; $c(1) \in M_1$, $\dot c(1) \in T^\perp_{c(1)}M_1$. In particular, if $M_0 \times M_1 = \{p\} \times \{q\}$, the critical points are the geodesics from p to q.*

(ii) *The critical points of $E \mid \Lambda M$ are the constant maps or the closed geodesics, i.e., $\dot c(0) = \dot c(1)$, $\nabla \dot c = 0$, $|\dot c| \neq 0$.*

PROOF. First assume $\partial c \in H^1(c^*TM)$. By partial integration we get

$$DE(c).\eta = \langle \partial c, \nabla\eta\rangle_0 = -\langle \nabla\partial c, \eta\rangle_0 + \langle \dot c(1), \eta(1)\rangle - \langle \dot c(0), \eta(0)\rangle.$$

Thus, $\nabla\partial c = 0$ and $\dot c(0) \perp T_{c(0)}M_0$, $\dot c(1) \perp T_{c(1)}M_1$ imply $DE(c).\eta = 0$ whenever $\eta \in T_c H^1_{M_0\times M_1}M$.

Similarly one sees that a closed geodesic or a constant map is a critical point for $E \mid \Lambda M$.

Conversely, let $DE(c) \mid T_c H^1_{M_0\times M_1}M = 0$. Determine H^1-vector fields $\zeta(t)$ and $\xi(t)$ along $c(t)$ by

$$\nabla\zeta(t) = \partial c(t) = \dot c(t), \quad \zeta(0) = 0; \qquad \nabla\xi(t) = 0, \quad \xi(1) = \zeta(1).$$

Put $\zeta(t) - t\xi(t) = \eta(t)$. Then $\eta(0) = \eta(1) = 0$; $\nabla\eta(t) = \partial c(t) - \xi(t)$. Moreover,

$$\langle \xi, \partial c - \xi\rangle_0 = \langle \xi, \nabla\eta\rangle_0 = \int_0^1 \frac{d}{dt}\langle \xi(t), \eta(t)\rangle\, dt = 0$$

and

$$0 = DE(c).\eta = \langle \partial c, \nabla\eta\rangle_0 = \langle \partial c, \partial c - \xi\rangle_0.$$

Hence, $\|\partial c - \xi\|_0^2 = 0$, i.e., $\partial c = \xi$ is of class H^1 and $\nabla\partial c = \nabla\xi = 0$; thus, c is a geodesic.

From the formula

$$DE(c).\eta = \langle \dot c(1), \eta(1)\rangle - \langle \dot c(0), \eta(0)\rangle = 0$$

we get $\dot c(1) \perp T_{c(1)}M_1$, $\dot c(0) \perp T_{c(0)}M_0$. In the case $c \in \Lambda M$, $c \neq$ const, we need to show $\dot c(0) = \dot c(1)$. But this follows from the same argument applied to $c(t + \frac{1}{2})$ instead of $c(t)$. □

We continue with some preliminary estimates.

2.4 PROPOSITION. *Let $d = d_M$ be the distance on M derived from the Riemannian metric on M. Let $c \in H^1(I, M)$. Then*

$$d(c(t_0), c(t_1)) \leq |t_1 - t_0|^{1/2}\sqrt{2E(c)}.$$

PROOF. We apply Cauchy-Schwarz:

$$d(c(t_0), c(t_1))^2 \leqslant \left(\int_{t_0}^{t_1} |\dot{c}(t)|\, dt\right)^2 \leqslant |t_1 - t_0|\, 2E(c). \quad \square$$

2.5 PROPOSITION. *Let c, c' in $H^1(I, M)$. Then $|\sqrt{2E(c)} - \sqrt{2E(c')}| \leqslant d_{H^1}(c, c')$, where d_{H^1} denotes the distance derived from the Riemannian metric on $H^1(I, M)$. This relation is true a fortiori if c and c' belong to some submanifold $H^1_N M$ and we replace d_{H^1} by the distance on $H^1_N M$.*

REMARK. In the case $M = \mathbf{R}^n$, 2.5 becomes the trivial estimate $|\|\partial c\|_0 - \|\partial c'\|_0| < \|c - c'\|_1$.

PROOF. We can assume that c, c' belong to the same connected component of $H^1_N M$. Then $d_{H^1}(c, c')$ is the infimum of the length $L(F)$ of curves $F: [0, 1] \to H^1(I, M)$ from $c = F(0)$ to $c' = F(1)$. An approximation argument shows that it suffices to consider the case $E(F(s)) > 0$ for all $s \in [0, 1]$. Putting $F(s)(t) = F(t, s)$ we find

$$\begin{aligned} \frac{d}{ds}\|\partial F\|_0 &= \frac{1}{\|\partial F\|_0}\int_I \left\langle \frac{\partial F}{\partial t}, \nabla \frac{\partial F}{\partial s}\right\rangle (t, s)\, dt \\ &\leqslant \left(\int_I \left|\nabla \frac{\partial F}{\partial s}(t, s)\right|^2 dt\right)^{1/2} \leqslant \left\|\frac{dF}{ds}(s)\right\|_1. \end{aligned}$$

Integration yields the claim. $\square$

2.6 LEMMA. *The inclusion $H^1_N M \to C^0_N M$ is continuous and also compact whenever one of the projections $\mathrm{pr}_1 N \subset M$, $\mathrm{pr}_2 N \subset M$ of $N \subset M \times M$ is compact. This is true in particular for ΩM, and it is also true for ΛM if M is compact.*

PROOF. The continuity is clear from the definition; cf. also 1.2. As for the compactness, consider a bounded sequence $\{c_m\}$ on $H^1_N M$. 2.5 then implies that $\{E(c_m)\}$ is bounded and 2.4 shows that $\{c_m\}$ is equicontinuous. The claim now follows if the evaluation set $\{c_m(t_0)\}$, for every $t_0 \in I$, is relatively compact. To see this, assume that $\mathrm{pr}_1 N$ is compact, i.e., $\{c_m(0)\}$ is relatively compact. 2.4 then shows that this is also true for $\{c_m(t_0)\}$. $\square$

2.7 THEOREM. *$H^1_N M$, with its distance deduced from the Riemannian metric, is a complete metric space provided $\mathrm{pr}_1 N \subset M$ or $\mathrm{pr}_2 N \subset M$ is compact. In particular, $M = \Omega_{pq} M$ and ΛM (for M compact) are complete metric spaces.*

PROOF. Let $\{c_m\}$ be a Cauchy sequence. From 2.6 we then know that there exists a limit $c_0 \in C^0_N M$ for this sequence. Since c_0 can be approximated by an H^1-curve c, we may assume that, for all large m, c_m is in a closed set contained in the domain $\mathfrak{U}(c)$ of the natural chart based at c. Appealing to the Riemann Principle (cf. Klingenberg [6]) we need only show that $\{\xi_m = \exp_c^{-1} c_m\} \in T_c \equiv T_c H^1_N M$ is convergent. But this is clear, since T_c is a closed linear subspace of the Hilbert space $T_c H^1(I, M) = H^1(c^* TM)$. $\square$

2.8 Definition. Let $H^1_N M$ be one of the submanifolds of $H^1(I, M)$ introduced in 2.2. Then we define the gradient vector grad $E(c)$ in $T_c H^1_N M$ as a representation of the 1-form $DE(c)$, i.e.,

$$\langle \operatorname{grad} E(c), \eta \rangle_1 = DE(c).\eta, \quad \text{all } \eta \in T_c H^1_N M.$$

Note. If c is differentiable and $H^1_N M = \Omega M$ or ΛM, we can apply partial integration and find that grad E is the solution, ξ, of $\nabla^2 \xi(t) - \xi(t) = \nabla \partial c(t)$ satisfying the boundary conditions $\xi(0) = \xi(1) = 0$ or $\xi(1) - \xi(0) = \nabla\xi(1) - \nabla\xi(0) = 0$.

We can now formulate *condition* (C) of Palais and Smale [**1**]:

'Let $\{c_m\}$ be a sequence such that $\{E(c_m)\}$ is bounded and $\{\|\operatorname{grad} E(c_m)\|_1\}$ is a null sequence. Then $\{c_m\}$ possesses accumulation points, and each such point c is a critical point, i.e., grad $E(c) = 0$.'

2.9 Theorem. *Condition* (C) *holds on* ΛM, *provided* M *is compact, and it always holds on* ΩM.

Note. Our proof will show that, more generally, condition (C) holds on $H^1_N M$, whenever either $\mathrm{pr}_1 N$ or $\mathrm{pr}_2 N$ is compact. See Grove [**1**].

Proof. We know from the proof of 2.7 that we can assume—by going to a subsequence, if necessary, which shall again be denoted by $\{c_m\}$—that all c_m belong to the domain $\mathcal{U}(c)$ of a natural chart and form a Cauchy sequence in the d_∞-metric. Put $\exp_c^{-1} c_m = \xi_m$. Invoking the Riemann Principle, cf. Klingenberg [**6**], we know that it suffices to show that $\{\xi_m\}$ is a Cauchy sequence in $T_c \equiv T_c H^1_N M$. We already know that $\{\xi_m\}$ is a Cauchy sequence in the norm $\|\ \|_\infty$ and hence also in the norm $\|\ \|_0$; cf. 1.5. The local representative on $\partial_c \xi_m$ of ∂c_m (see 1.16) tells us that $\{\partial_c \xi_m\}$ is a Cauchy sequence in the $\|\ \|_0$-norm if and only if this is the case for the sequence $\{\nabla \xi_m\}$.

We first show that $\{\|\xi_m\|_1\}$ is bounded. Indeed, from the Riemann Principle we find that $\|\partial_c \xi_m\|_0^2$ is of a size comparable to $\langle \partial c_m, \partial c_m \rangle_0 = 2E(c_m)$ with a factor independent of m. By looking at formula 1.10 for $DE(c)$ we see by the same argument that

$$\langle \partial_c \xi_l, \partial_c \xi_l - \partial_c \xi_m \rangle_0 = DE(c_l).(\xi_l - \xi_m),$$

modulo terms which go to 0, as $l, m \to \infty$. Here we have interpreted ξ_l, ξ_m as elements in $T_{c_l} H^1_N M$, which has in our chart the representative $T_c H^1_N M$. We now write, modulo such vanishing terms,

$$\begin{aligned}\|\partial_c \xi_l - \partial_c \xi_m\|_0^2 &= \langle \partial_c \xi_l, \partial_c \xi_m - \partial_c \xi_m \rangle_0 + \langle \partial_c \xi_m, \partial_c \xi_m - \partial_c \xi_l \rangle_0 \\ &= DE(c_l).(\xi_l - \xi_m) + DE(c_m).(\xi_m - \xi_l).\end{aligned}$$

But

$$|DE(c_m).(\xi_l - \xi_m)| \leq \|\operatorname{grad} E(c_m)\|_1 \|\xi_l - \xi_m\|_1,$$

and $\|\text{grad } E(c_m)\|_1 \to 0$ for $m \to \infty$, while $\|\xi_l - \xi_m\|_1$ remains bounded. Therefore, $\|\partial_c\xi_l - \partial_c\xi_m\|_0 \to 0$ for $l, m \to \infty$. □

There are a few immediate consequences of the validity of condition (C). First we introduce the following notation:

2.10 DEFINITION. (i) As joint notation for the spaces $\Omega = \Omega M = \Omega_{pq}M$ and ΛM (for M compact) we use P.

(ii) For any real number κ we define

$$P^\kappa = \{c \in P;\ E(c) \leqslant \kappa\} \quad \text{and} \quad P^{\kappa-} = \{c \in P;\ E(c) < \kappa\}.$$

Note. For $P = \Omega_{pq}M$, $\kappa_0 = d(p,q)^2/2$ is the smallest value of $E \mid P$. Hence, for $\kappa < \kappa_0$, P^κ is empty. For $P = \Lambda M$, $\inf E \mid \Lambda M = 0$.

2.11 PROPOSITION. *Denote by* $\operatorname{Cr} P$ *the set of critical points on* P. *Choose a real number* κ. *Put* $\operatorname{Cr} P \cap E^{-1}(\kappa) = \operatorname{Cr} \kappa$. *Then* $\operatorname{Cr} \kappa$ *is compact. The same is true of* $\operatorname{Cr} P^\kappa = \operatorname{Cr} P \cap P^\kappa$.

PROOF. Immediate from 2.9. □

2.12 LEMMA. *Choose an open neighborhood* $\mathfrak{U}$ *of* $\operatorname{Cr}\kappa$ *in* P. *Then there exist* ε, $\eta > 0$ *such that*

$$c \in \complement\, \mathfrak{U} \cap (P^{\kappa+\varepsilon} - P^{\kappa-\varepsilon}) \textit{ implies } \|\text{grad } E(c)\|_1 \geqslant \eta.$$

PROOF. Let $\{c_m\}$ be a sequence with $\{E(c_m)\}$ converging towards κ and $\{\|\text{grad } E(c_m)\|_1\}$ going to zero. Condition (C) implies that the elements c_m, for all sufficiently large m, will lie in the prescribed neighborhood $\mathfrak{U}$. □

2.13 DEFINITION. The integral curve of the vector field $-\text{grad } E$ on P which starts for $s = 0$ at c is denoted by $\phi_s c$.

REMARKS. One knows from the theory of differential equations that for every $c \in P$ there exists a maximal interval $J = J(c)$ containing $0 \in \mathbf{R}$ on which $\phi_s c$, $s \in J$, is defined (cf. Dieudonné [**1**]). If grad $E(c) = 0$, $\phi_s c = c$ for all $s \in \mathbf{R}$, i.e., in this case $J(c) = \mathbf{R}$. As we will show in 2.15, $J(c) = \mathbf{R}$ for all other $c \in P$ also.

2.14 LEMMA.

(i) $$\frac{d}{ds}E(\phi_s c) = DE(\phi_s c).(-\text{grad } E(\phi_s c)) = -\|\text{grad } E(\phi_s c)\|_1^2 \leqslant 0.$$

Hence, for $s_0 \leqslant s_1$, $E(\phi_{s_1}c) - E(\phi_{s_0}c) \leqslant 0$.

(ii) *Denote by* $d_P(\ ,\ )$ *the distance on* P *derived from the Riemannian metric. Then, if* $s_0 \leqslant s_1$,

$$d_P(\phi_{s_1}c, \phi_{s_0}c)^2 \leqslant \left(\int_{s_0}^{s_1}\left\|\frac{d\phi_s c}{ds}\right\|ds\right)^2$$
$$\leqslant \left|E(\phi_{s_1}c) - E(\phi_{s_0}c)\right| |s_1 - s_0| \leqslant E(\phi_{s_0}c)\,|s_1 - s_0|.$$

(iii) $\|\text{grad } E(c)\|_1^2 \leqslant 2E(c)$.

(iv) $E(\phi_s c) \leqslant E(c) + E(c)e^{-2s}$ for all $s \in J(c)$.

PROOF. (i) is an immediate consequence of the definitions. In (ii) we use the Schwarz inequality. As for (iii), we have

$$\|\text{grad } E(c)\|_1^2 = DE(c).\text{grad } E(c) = \langle \partial c, \nabla \text{ grad } E(c)\rangle_0$$
$$\leq \|\partial c\|_0 \|\nabla \text{ grad } E(c)\|_0 \leq \sqrt{2E(c)}\,\|\text{grad } E(c)\|_1.$$

To prove (iv) we have from (i) and (iii), for $E(c) > 0$,

$$\frac{d}{ds}E(\phi_s c) \geq -2E(\phi_s c); \quad \frac{d}{ds}\ln E(\phi_s c) \geq -2.$$

Therefore, if $s \leq 0$, $E(\phi_s c) \leq E(c)e^{-2s}$. From this we get, for all $s \in J(c)$,

$$E(\phi_s c) \leq \max(E(c), E(c)e^{-2s}) \leq E(c) + E(c)e^{-2s}. \quad \square$$

2.15 THEOREM. *The integral curve $\phi_s c$ is defined for all $s \in \mathbf{R}$.*

Note. The fact that $\phi_s c$ is defined not only for $s > 0$ (when it is a simple consequence of $E(\phi_s c) > 0$) was observed by Solà-Morales [**1**].

PROOF. But $J(c) =]s_-, s_+[$. If $s_- > -\infty$ or $s_+ < +\infty$, there exists a Cauchy sequence $\{s_m\}$, $s_- < s_m < s_+$, with limit a finite boundary point of $J(c)$. From 2.14(iv), (ii) it follows that $\{\phi_{s_m} c\}$ is a Cauchy sequence. Let $\tilde{c}$ be its limit. There exists a neighborhood $\mathcal{U}$ of $\tilde{c}$ and an $\varepsilon > 0$ such that $\phi_s c^*$ is defined for all $c^* \in \mathcal{U}$, all $|s| < 2\varepsilon$. In particular, $\phi_s \phi_{s_m} c = \phi_{s+s_m} c$ is defined for $|s| = \varepsilon$, m large. But $\varepsilon + s_m$ or $-\varepsilon + s_m$ lies outside $J(c)$ for large m, a contradiction. $\square$

2.16 LEMMA. *Let κ be a noncritical value of E. Then there exist ε, $s_0 > 0$ such that $\phi_s P^{\kappa+\varepsilon} \subset P^{\kappa-\varepsilon}$, all $s \geq s_0$.*

REMARK. This lemma will be fundamental for the subsequent existence proofs of critical points.

PROOF. From 2.12 we have, with $\mathcal{U} = \varnothing$, an $\eta > 0$ such that $\|\text{grad } E(c)\| > \eta$ for c with $\kappa - \varepsilon < E(c) < \kappa + \varepsilon$. Put $2\varepsilon/\eta^2 = s_0$. Whenever $c \in P^{\kappa-\varepsilon}$ then $\phi_s c \in P^{\kappa-\varepsilon}$ for $s \geq 0$. Therefore it remains to consider c's with $\kappa - \varepsilon < E(c) \leq \kappa + \varepsilon$. We derive a contradiction from the assumption that $E(\phi_s c) > \kappa - \varepsilon$ for $0 \leq s \leq s_0$. Indeed, if this were the case, we would get, from 2.14,

$$E(\phi_{s_0} c) = E(c) - \int_0^{s_0} \|\text{grad } E(\phi_s c)\|_1^2\, ds \leq \kappa + \varepsilon - \eta^2 s_0 = \kappa - \varepsilon. \quad \square$$

2.17 DEFINITION. Let κ_0 be such that there are no critical values in $]\kappa_0, \kappa_0 + \varepsilon[$ for some small $\varepsilon > 0$. A ϕ-family of P mod P^{κ_0} is a family $\mathcal{A}$ of nonempty subsets A of P such that:

(i) $E \mid A$ is bounded;

(ii) $A \in \mathcal{A}$ implies $\phi_s A \in \mathcal{A}$ for $s \geq 0$;

(iii) $\sup E \mid A \geq \kappa_0 + \varepsilon$.

The critical value of such a family $\mathcal{A}$ is defined by

$$\kappa_{\mathcal{A}} = \inf_{A \in \mathcal{A}} \sup E \mid A.$$

In the case $\kappa_0 < \inf E \mid P$, we simply speak of a ϕ-family of P.

EXAMPLES. (i) Consider the elements $\{\phi_s c\}$ of a ϕ-trajectory, all $s \geqslant 0$. Let $\lim_{s\to\infty} E(\phi_s c) = \kappa^*$. Then this is a ϕ-family of P mod P^{κ_0} for any $\kappa_0 < \kappa^*$. In the same way one gets from A, with $E\,|\,A$ bounded, the ϕ-family $\{\phi_s A;\ \text{all } s \geqslant 0\}$.

(ii) Let P' be a connected component of P. Then the $c \in P'$ form a ϕ-family.

(iii) Consider $P = \Lambda M$. $\kappa_0 = 0$ is a critical value such that there are no critical values in $]\kappa_0, \kappa_0 + \varepsilon] =]0, \varepsilon]$, $\varepsilon > 0$ sufficiently small. Indeed, take some ε, $0 < \varepsilon < 2\iota(M)^2$, where $\iota(M) \leqslant$ the injectivity radius of the compact manifold M. Then $E(c) < \varepsilon$ implies $L(c) < \sqrt{2E(c)} < 2\iota(M)$. Thus, for $c \in \Lambda^\varepsilon M$, $c(t)$ belongs to the ball $B_{\iota(M)}(c(0))$ around $c(0)$. c cannot be a closed geodesic since then $c(\frac{1}{2}) = \exp \frac{1}{2}\dot{c}(0)$ must be equal to $\exp(-\frac{1}{2}\dot{c}(0))$. Thus, $0 = \inf E\,|\,\Lambda M$ is always an isolated critical value.

While $\kappa_0 = \inf E\,|\,\Omega M$ certainly is a critical value, it need not be isolated.

(iv) Let $F: (\bar{B}^k, \partial\bar{B}^k) \to (\Lambda M, \Lambda^0 M)$ be homotopically nontrivial. Let $\varepsilon > 0$ be as in (iii). Then there can be no F' homotopic to F with im $F' \subset \Lambda^\varepsilon M$. Indeed, we could then deform each $F'(x)$ into $F'(x)(0) \in \Lambda^0 M$, and this simultaneously for all $x \in \bar{B}^k$. Therefore, the set $\{F'(\bar{B}^k);\ F' \text{ homotopic to } F\}$ forms a ϕ-family of ΛM mod $\Lambda^0 M$.

(v) Let $w \in H_*(P, P^{\kappa_0})$ be a nontrivial homology class. Assume that $P^{\kappa_0+\varepsilon'}$ can be deformed into some subset of P^{κ_0} for all sufficiently small ε'. We obtain a ϕ-family P mod P^{κ_0} from w by taking as its sets the images of the singular simplices belonging to the cycles u representing w.

The importance of the concept of a ϕ-family lies in the fact that its critical value is a critical value of E.

2.18 THEOREM. *The critical value $\kappa = \kappa_{\mathfrak{A}}$ of a ϕ-family $\mathfrak{A}$ of P mod P^{κ_0} is $> \kappa_0$ and it is a critical value of E. Assume that the members A of the family $\mathfrak{A}$ are compact. Let $\mathfrak{U}$ be an open neighborhood of the set* Cr κ *of critical points with E-value $\kappa = \kappa_{\mathfrak{A}}$. Then there exists an $A' \in \mathfrak{A}$ such that $\phi_s A' \subset \mathfrak{U} \cap P^{\kappa-}$ for all $s \geqslant 0$.*

Note. One occasionally describes this by saying 'A' remains hanging at Cr κ'.

PROOF. The definition of $\kappa = \kappa_{\mathfrak{A}}$ implies that for every $\varepsilon > 0$ there is an $A \in \mathfrak{A}$ with $\sup E\,|\,A < \kappa + \varepsilon$. Thus, $\kappa > \kappa_0$. If κ were noncritical, 2.16 would yield an $\varepsilon > 0$ and $s_0 > 0$ with $\sup E\,|\,\phi_{s_0} A < \kappa - \varepsilon$ with $A \in \mathfrak{A}$ as above. This proves the first statement.

Now fix an open neighborhood $\mathfrak{U}$ of Cr κ. Let $\mathfrak{V}$, $\mathfrak{W}$ be open neighborhoods of Cr κ satisfying $\mathfrak{U} \supset \mathfrak{V} \supset \mathfrak{W}$ and such that there exists a $\rho > 0$ with $d_P(c', c'') > \rho$, for $c' \in \mathfrak{V}$, $c'' \notin \mathfrak{U}$; $d_P(c, c') > \rho$, for $c \in \mathfrak{W}$, $c' \notin \mathfrak{V}$. That there exist such neighborhoods follows from the fact that Cr κ is compact and d_P is a distance on P.

According to 2.12 there exist ε, $\eta > 0$ such that $\|\text{grad } E(c)\|_1 \geqslant \eta$ for $c \notin \mathfrak{W}$ and $\kappa - \varepsilon \leqslant E(c) \leqslant \kappa + \varepsilon$. Let $\phi_s c$, $0 \leqslant s \leqslant s_1$, be a trajectory from a point

$c \in \mathcal{W}$, with $\kappa - \varepsilon \leqslant E(c) \leqslant \kappa + \varepsilon$, to a point $\phi_{s_1} c = c' \notin \mathcal{V}$, with $\kappa - \varepsilon \leqslant E(\phi_s c) \leqslant \kappa + \varepsilon$. Since this trajectory contains an arc of length $> \rho$ in the set

$$(*) \qquad \mathcal{W} \cap (P^{\kappa+\varepsilon} - P^{(\kappa-\varepsilon)-}),$$

we can estimate the decrease of E along this trajectory by

$$E(c') - E(c) = -\int_0^{s_1} \|\text{grad } E(\phi_s c)\|_1^2 \, ds \leqslant -\eta^2 s_1.$$

From 2.14(ii) we have $\rho^2 \leqslant E(c)s_1 \leqslant (\kappa + \varepsilon)s_1$, i.e.,

$$E(c') \leqslant E(c) - \eta^2\rho^2/(\kappa + \varepsilon) < E(c) - \eta^2\rho^2/\kappa.$$

Hence, by taking $\varepsilon > 0$ sufficiently small, we can assume that the trajectory $\phi_s c$, $s \geqslant 0$, of an element $c \in \mathcal{W} \cap P^{\kappa+\varepsilon-}$ remains in the set $\mathcal{V} \cup P^{(\kappa-\varepsilon)-}$ because when it leaves $\mathcal{V}$ it has lost an amount $> 2\varepsilon$ of its E-value and therefore has entered $P^{(\kappa-\varepsilon)-}$.

For the same reason we can assume that the trajectory $\phi_s c'$, $s > 0$, of an element $c' \in \mathcal{V} \cup P^{\kappa+\varepsilon}$ remains in $\mathcal{V} \cup P^{(\kappa-\varepsilon)-}$.

Now choose $A \in \mathcal{Q}$, $A \subset P^{\kappa+\varepsilon}$. For every $c \in A$ there is an $s_0 = s(c) \geqslant 0$ such that $\phi_s c$, $s \geqslant s_0$, lies in $\mathcal{V} \cup P^{(\kappa-\varepsilon)-}$. Indeed, since $\|\text{grad } E\|_1$ is bounded away from zero on the set $(*)$, there will be an $s_1 > 0$ with $\phi_{s_1} c \in \mathcal{W} \cap P^{(\kappa+\varepsilon)-}$, and we can apply our previous result. With such an $s_0 = s(c)$ we can find an open neighborhood $\mathcal{U}(c)$ of c such that $\phi_{s_0}\mathcal{U}(c) \subset \mathcal{V} \cup P^{(\kappa-\varepsilon)-}$. As we have seen above, we then have $\phi_s \mathcal{U}(c) \subset \mathcal{U} \cup P^{(\kappa-\varepsilon)-}$ for all $s \geqslant s_0$. Since A is compact, a finite number of such $\mathcal{U}(c)$ cover A. Therefore there will be an $s_0 \geqslant 0$ such that $\phi_s A \subset \mathcal{U} \cup P^{\kappa-}$, all $s \geqslant s_0$. Put $\phi_{s_0} A = A'$. □

As a first consequence we mention the

2.19 **Theorem.** *On every connected component P' of P, E assumes its infimum. If c is an element in P' with minimal E-value then c is critical. That is, in the case $P = \Omega_{pq}M$, c is a geodesic from p to q. In the case $P = \Lambda M$, c is a constant map if $\Lambda' M$ consists of the nullhomotopic curves; otherwise c is a closed geodesic.* □

Remarks with examples. The connected components of $\Omega M = \Omega_{pq}M$ are in 1 : 1 correspondence with the elements of the fundamental group $\pi_1 M$ of M. The connected components of ΛM correspond to the conjugacy classes of $\pi_1 M$. Thus, if we take, e.g., for M a torus of revolution in Euclidean 3-space, there are at least as many geodesics from a point p to a point q as there are elements in $\pi_1 M = \mathbf{Z} \times \mathbf{Z}$. Since $\pi_1 M$ is abelian, the same is true for the number of closed geodesics. Actually, in each class of freely homotopic, not null homotopic closed curves on a torus, there are at least two closed geodesics not just differing by their parameterizations. One is of minimal length in its class, the other of minimax type, i.e., with index plus nullity equal to one. This is due to the fact that a connected component of $\Lambda(S^1 \times S^1)$, when we divide out by the parameterizations, has the homotopy type of S^1.

In the case of a manifold of negative sectional curvature, each connected component of ΛM not containing the trivial curves has, modulo parameterization, the homotopy type of a point. This is the reason for the uniqueness of a curve of minimal length (up to parameterization) in such a homotopy class.

The previous theorem does not yield the existence of a closed geodesic in case the manifold is simply connected. Appealing to some well-known facts in the topology of manifolds we can show

2.20 THEOREM. *On any compact Riemannian manifold there exists a closed geodesic.*

REMARK. This theorem was proved by Lyusternik and Fet [**1**]. Our proof is somewhat different than the original one.

PROOF. If $\pi_1 M \neq 0$, the theorem is contained in 2.19. So assume $\pi_1 M = 0$. Since M is compact, there exists k, $0 \leqslant k < \dim M$, such that the homotopy group $\pi_{k+1} M$ of M is nontrivial. That is, there exists a

$$(*) \qquad f: S^{k+1} \to M$$

which is not homotopic to a constant mapping (cf. Spanier [**1**]). We associate with such an f a mapping

$$(**) \qquad F = F(f): (\bar{B}^k, \partial \bar{B}^k) \to (\Lambda M, \Lambda^0 M)$$

as follows: First, identify the closed k-ball $\bar{B}^k = \{x \in \mathbf{R}^k; |x| \leqslant 1\}$ with the half equator on $S^{k+1} \subset \mathbf{R}^{k+2}$ given by $\{x = (x_0, \ldots, x_{k+1}) \in S^{k+1}; x_0 \geqslant 0$ and $x_1 = 0\}$. Denote by $c_p(t)$, $0 \leqslant t \leqslant 1$, the circle which starts out from $p \in \bar{B}^k$ orthogonally to the hyperplane $\{x_1 = 0\}$ and enters the half sphere $\{x_1 > 0\}$. If $p \in \partial \bar{B}^k$, c_p is in the trivial circle $c_p(t) = p$, of course. With this we put

$$F(p) = \{f \circ c_p(t); 0 \leqslant t \leqslant 1\}.$$

Note that, conversely, a map F of $(\bar{B}^k, \partial \bar{B}^k)$ into $(\Lambda M, \Lambda^0 M)$ determines a map $f = f(F): S^{k+1} \to M$ such that $F(f(F)) = F$ and $f(F(f)) = f$. Indeed, describe $q \in S^{k+1}$ by $c_p(t)$. t is uniquely determined, except when $q \in \bar{B}^k \subset S^{k+1}$, in which case $q = c_q(t) = \text{const}$. Now define $f(q)$ by $F(p)(t)$.

Even more is true. If f_s, $0 \leqslant s \leqslant 1$, is a homotopy of $f = f_0$, $(*)$, then the associated family $F_s = F(f_s)$, $(**)$, is a homotopy of $F = F_0$. And a homotopy F_s of $F = F_0$ determines a homotopy $f_s = f(F_s)$ of $f(F) = f(F_0)$.

Recall from example (iii) after 2.17 that the family of sets $F'(\bar{B}^k)$, F' homotopic to F, is a ϕ-family of ΛM mod $\Lambda^0 M$. □

One can ask whether there exists more than one closed geodesic on a compact Riemannian manifold M. Of course, two closed geodesics which differ only by the choice of the initial point or the orientation should, in this connection, not be counted as different. Also, closed geodesics which are just different coverings of an underlying prime closed geodesic should not be counted as geometrically different. See the Appendix and in particular the remark there at the end for a precise formulation of what it means that two geodesics are geometrically different.

In our present exposition of Riemannian geometry we will give only a few results concerning the existence of geometrically different geodesics. For a full account we must refer the reader to Klingenberg [**4, 5**]. One of the main results there is the existence of infinitely many geometrically distinct closed geodesics, provided M has finite fundamental group.

The corresponding question for the loop space $\Omega = \Omega_{pq}$ of curves from p to q is much easier to answer. Serre [**1**] has shown that there exist infinitely many geodesics from p to q if and only if some homotopy group $\pi_k M$, $k > 1$, of M is nonzero. For M compact, this is known to always be true. What Serre shows is that $\pi_k M \neq 0$, for some $k > 1$, implies that the sequence $\{b_i \Omega M\}$ of Betti numbers of ΩM contains infinitely many nonzero elements. This then yields the existence of infinitely many critical points in ΩM.

A note of caution must be added. As shown by the sphere with the standard metric, it can happen that the infinitely many geodesics from p to q all lie on the same closed geodesic. Thus one might ask whether it is really justified to speak of infinitely many geometrically different geodesics from p to q. However, such a phenomenon presumably can happen only for very special Riemannian metrics: In general, for almost all p and q on M, different geodesics from p to q have no arcs in common.

Chapter 3. The second order neighborhood of a critical point

We continue to consider the functional E on one of the Hilbert submanifolds of $H^1(I, M)$ which satisy condition (C). Our main interest is again directed towards the loop space $\Omega M = \Omega_{pq}M$ and the space ΛM of closed curves—sometimes also called a free loop space.

We study the second order neighborhood of a critical point c of E, i.e., the Hessian $D^2E(c)$. For the cases mentioned above, $D^2E(c)$ has a particularly simple expression, involving the curvature tensor only (cf. 3.1). The associated selfadjoint operator A_c is of the form identity plus compact operator (cf. 3.2). In particular the negative eigenspace of A_c has finite dimension; it is called the index of c; cf. 3.3, 3.6. The null space of A_c consists of the Jacobi fields which belong to the tangent space at c to the submanifold under consideration (cf. 3.6). In 3.7 we determine the eigenspaces of A_c for c a critical point of ΛS^n.

The Morse Index Theorem (3.9) for a critical point c of $\Omega_{pq}M$ relates the index of c to the number of conjugate points c. 3.14 is the analogue for a closed geodesic.

We conclude with an important generalization, 3.15, of 2.16 where the index of the critical points comes into play. Usually, this result is proved by first considering the case where all critical points are nondegenerate (this is the hypothesis for the so-called Morse theory proper) and then employing an approximation argument. Our direct proof uses a novel technique. The result may be viewed as belonging to the common realm of Lusternik-Schnirelmann and Morse theory.

In the Appendix we describe the natural S^1-action and $\mathbf{Z}_2$-action on the manifold ΛM of closed H^1-curves.

We continue to consider the functional E on one of the Hilbert manifolds $H^1_{M_0\times M_1}M$ or ΛM of $H^1(I, M)$. In the first case we assume M_0 or M_1 to be compact. In the second case, M shall be compact. Thus, condition (C) holds in all cases.

We begin by defining a linear operator with the help of the curvature tensor. Let c be a geodesic. We start with the linear bundle mapping

$$R_c\colon c^*TM \to c^*TM;\quad \xi_t \in (c^*\tau)^{-1}(t) \mapsto R_{\dot c(t)}\xi_t \in (c^*\tau)^{-1}(t).$$

Here, $R_{\dot c(t)}\xi_t = R(\xi_t, \dot c(t))\dot c(t)$ is the curvature operator (cf. Klingenberg [**6**]). We consider the induced mapping (cf. 1.6)

$$\tilde R_c\colon H^1(c^*TM) \cong T_cH^1(I, M) \to T_cH^1(I, M).$$

Note that $\xi_0 = \xi_1 = 0$ implies $R_c\xi_0 = R_c\xi_1 = 0$, and $\xi_0 = \xi_1$ implies $R_c\xi_0 = R_c\xi_1$ if $\dot{c}(0) = \dot{c}(1)$. Thus, $\tilde{R}_c$ transforms the subspace $T_c\Omega_{c(0)c(1)}M$ into itself and, if c is a closed geodesic, then $T_c\Lambda M$ is also transformed into itself.

We compute $D^2E(c)$:

3.1 LEMMA. (i) *Let c be a critical point of $E \mid H^1_{M_0\times M_1}M$. Let ξ, ξ' be elements of $T_cH^1_{M_0\times M_1}M$. Then*

$$(*) \qquad D^2E(c)(\xi,\xi') = \langle \nabla\xi, \nabla\xi'\rangle_0 - \langle \tilde{R}_c\xi, \xi'\rangle_0 + h_{\dot{c}(1)}(\xi(1), \xi'(1)) - h_{\dot{c}(0)}(\xi(0), \xi'(0)).$$

Here, $h_{\dot{c}(1)}(\ ,\ )$ is the second fundamental form of M_i at $c(i)$ in the direction of the normal vector $\dot{c}(i)$, $i = 0, 1$.

(ii) *If M_i is totally geodesic at $c(i)$, $i = 0, 1$, $(*)$ reduces to*

$$(**) \qquad D^2E(c)(\xi,\xi') = \langle \nabla\xi, \nabla\xi'\rangle_0 - \langle \tilde{R}_c\xi, \xi'\rangle_0.$$

This holds, in particular, for $\Omega M = \Omega_{c(0)c(1)}M$.

(iii) $(**)$ *also holds for c a critical point of $E \mid \Lambda M$.*

PROOF. It suffices to take $\xi = \xi'$ and then apply polarization. Let c_s, $s \in]-\varepsilon, \varepsilon[$ be a curve on $H^1_{M_0\times M_1}M$ with $c_0 = c$ and $dc_s/ds\,|_0 = \xi$. Put $c_s(t) = F(s,t)$. Then

$$\frac{d}{ds}E(c_s) = \int_0^1 \left\langle \frac{\nabla}{\partial s}\frac{\partial}{\partial t}F, \frac{\partial F}{\partial t}\right\rangle(s,t)\,dt = \int_0^1 \left\langle \frac{\nabla}{\partial t}\frac{\partial}{\partial s}F, \frac{\partial F}{\partial t}\right\rangle(s,t)\,dt,$$

$$(\S) \qquad \begin{aligned} D^2E(c)(\xi,\xi) &= \int_0^1 \left\langle \frac{\nabla}{\partial t}\frac{\partial}{\partial s}F, \frac{\nabla}{\partial t}\frac{\partial}{\partial s}F\right\rangle(0,t)\,dt \\ &\quad + \int_0^1 \left\langle R\left(\frac{\partial F}{\partial s}, \frac{\partial F}{\partial t}\right)\frac{\partial F}{\partial s}, \frac{\partial F}{\partial t}\right\rangle(0,t)\,dt \\ &\quad + \int_0^1 \frac{d}{dt}\left\langle \frac{\nabla}{\partial s}\frac{\partial}{\partial s}F, \frac{\partial F}{\partial t}\right\rangle(0,t)\,dt. \end{aligned}$$

The last term in (§), when integrated, is of the form $h_{\dot{c}(1)}(\xi(1), \xi(1)) - h_{\dot{c}(0)}(\xi(0), \xi(0))$. It vanishes for M_i totally geodesic at $c(i)$, $i = 0, 1$.

The same arguments apply to a critical point c of ΛM, the only difference being that this time the last term in (§) vanishes due to $F(s,1) = F(s,0)$; $\partial F(0,1)/\partial t = \partial F(0,0)/\partial t$. □

We analyse the structure of the operator associated to $D^2E(c)$ in the case ΩM and ΛM and $H^1_{M_0\times\{p_1\}}$ with M_0 of codimension 1 and totally geodesic at $p_0 = c(0)$. As common notation for these three cases we use P.

3.2 LEMMA. *Let c be a critical point of $E \mid P$. Then the selfadjoint operator A_c defined by*

$$\langle A_c\xi, \xi'\rangle_1 = \langle \xi, A_c\xi'\rangle_1 = D^2E(c)(\xi,\xi')$$

is of the form $A_c = 1 + k_c$. Here 1 is the identity and k_c is the compact operator given by $k_c = -(1-\nabla^2)^{-1}\circ(\tilde{R}_c + 1)$.

PROOF. Partial integration and the boundary conditions for differentiable $\eta, \eta' \in T_cP$ give the relation

$$\langle \eta, \eta' \rangle_1 = \langle \nabla\eta, \nabla\eta' \rangle_0 + \langle \eta, \eta' \rangle_0 = \langle -\nabla^2\eta, \eta' \rangle_0 + \langle \eta, \eta' \rangle_0 = \langle (1 - \nabla^2)\eta, \eta' \rangle_0.$$

By continuity, this relation extends to all of T_cP. Thus, for ξ, ξ' in T_cP,

$$\langle A_c\xi, \xi' \rangle_1 = \langle \nabla\xi, \nabla\xi' \rangle_0 - \langle \tilde{R}_c\xi, \xi' \rangle_0 = \langle \xi, \xi' \rangle_1 - \langle (\tilde{R}_c + 1)\xi, \xi' \rangle_0 = \langle \xi, \xi' \rangle_1 - \langle (1 - \nabla^2)^{-1} \circ (\tilde{R}_c + 1)\xi, \xi' \rangle_1.$$

As for the compactness of k_c we can either refer to general facts about the compactness of the inverse of an elliptic differential operator such as $-(1 - \nabla^2)$, or we can derive it directly as follows: From $\langle k_c\xi, k_c\xi \rangle_1 = -\langle (\tilde{R}_c + 1)\xi, k_c\xi \rangle_0$ we get

$$\| k_c\xi \|_1^2 \leq \| \tilde{R}_c + 1 \|_\infty \cdot \| k_c\xi \|_\infty \cdot \| \xi \|_0 \leq \text{const} \| k_c\xi \|_1 \cdot \| \xi \|_0.$$

We know from 2.6 that a bounded H^1-sequence $\{\xi_m\}$ is relatively compact as a C^0-sequence and hence as an H^0-sequence. Since $\|k_c\xi_m\|_1 \leq \text{const}\|\xi_m\|_0$, $\{k_c\xi_m\}$ is a relatively compact H^1-sequence. □

3.3 COROLLARY. *The operator A_c defined in* 3.2 *has either finitely many eigenvalues, including* 1, *or an infinite sequence of eigenvalues* = 1 *which have* 1 *as the only accumulation point.* 1 *is a spectral value, not an eigenvalue. In particular, the tangent space $T_c = T_c\Omega$ or $= T_c\Lambda M$ or $= T_cH^1_{M_0\times\{p_1\}}$ possesses an orthogonal decomposition*

$$T_c = T_c^- \oplus T_c^0 \oplus T_c^+$$

into a finite-dimensional subspace T_c^- spanned by the eigenvectors with negative eigenvalue, a finite-dimensional subspace T_c^0 spanned by the eigenvectors with eigenvalue zero, and a proper Hilbert space T_c^+ spanned by the eigenvectors with positive eigenvalue.

PROOF. This is an immediate consequence of the well-known structure of the spectrum of a compact operator (cf. Dieudonné [**1**]). □

3.4 DEFINITION. Let c be a critical point of $E \mid P$.

(i) $\dim T_c^-$ is called the index of c. According to the three cases, ΩM, ΛM and $H^1_{M_0\times\{p_1\}}$, we also speak of the Ω-index of c, $\text{index}_\Omega\, c$, the Λ-index of c, $\text{index}_\Lambda\, c$, and the focal index of c, $\text{index}_{\text{foc}}\, c$.

(ii) $\dim T_c^0$ is called the nullity of c, except in the case when $P = \Lambda M$ and $c_0 \neq \text{const}$. In that case the nullity of c is defined to be $\dim T_c^0 - 1$.

REMARKS. 1. The reason we define the nullity of a closed geodesic c to be $\dim T_c^0 - 1$ is that the vector field ∂c always belongs to T_c^0. This follows at once from 3.1 since $\tilde{R}_c\partial c \equiv 0$. Cf. also 3.6 below.

2. A trivial, but useful, observation is the following: If U is a subspace of $T_c = T_c\Omega M$ or $T_c\Lambda M$ or $T_cH^1_{M_0\times\{p_1\}}$ such that $D^2E(c) \mid U$ is negative semidefinite,

i.e., $\xi \in U$ implies $D^2E(c)(\xi, \xi) \leqslant 0$, then $\dim U \leqslant \operatorname{index} c + \dim T_c^0$. If, moreover, $U \cap T_c^0 = 0$, then $\dim U \leqslant \operatorname{index} c$.

This is a simple fact of linear algebra: Let

$$\mathrm{pr}_- : T_c = T_c^- \oplus T_c^0 \oplus T_c^+ \to T_c^-$$

be the orthogonal projection. Consider $\mathrm{pr}_-| U : U \to T_c^-$. Then $\xi \in \ker \mathrm{pr}_-| U$ must belong to $T_c^0 \oplus T_c^+$. That is, since $D^2E(c)(\xi, \xi) \leqslant 0$, $\xi \in T_c^0$. From

$$\dim U = \dim(\mathrm{image}\, \mathrm{pr}_-| U) + \dim(\mathrm{kernel}\, \mathrm{pr}_-| U),$$

we prove our claim.

3.5 PROPOSITION. *Let ξ be a eigenvector to an eigenvalue $\lambda \neq 1$ of A_c. Then ξ is a differentiable solution of*

$$\nabla^2\xi(t) + \frac{1}{1-\lambda} R(\xi, \dot{c})\dot{c}(t) + \frac{\lambda}{1-\lambda}\xi(t) = 0$$

satisfying the boundary condition $\xi \in T_c\Omega M$, $T_c\Lambda M$ or $T_c H^1_{M_0 \times \{p_1\}}$.

PROOF. In $A_c\xi = \lambda\xi$ substitute the expression $A_c = 1 + k_c$ with k_c as in 3.2.

3.6 COROLLARY. $\xi \in T_c^0$ *if and only if $\xi(t)$ is a differentiable Jacobi field along $c(t)$ satisfying the boundary condition $\xi \in T_c\Omega M$, $T_c\Lambda M$ or $T_c H^1_{M_0 \times \{p_1\}}$.*

Note. $\xi \in T_c\Lambda M$ and ξ differentiable means $(\xi(1), \nabla\xi(1)) = (\xi(0), \nabla\xi(0))$. Hence T_c^0 can be identified with the space of the periodic Jacobi fields along c.

PROOF. Put $\lambda = 0$ in 3.5. □

3.7 EXAMPLES. (i) Let M be flat, i.e., $\tilde{R}_c = 0$. Then the eigenvalue equation 3.5 reads $\nabla^2\xi + \lambda(1-\lambda)^{-1}\xi = 0$. For $c \in \Omega M$ or ΛM the boundary conditions $\xi(0) = \xi(1) = 0$ or $\xi(0) = \xi(1)$, $\nabla\xi(0) = \nabla\xi(1)$ allow no nontrivial solution for $\lambda < 0$. Thus, $\operatorname{index} c = 0$. On the other hand, for $\lambda = 0$, $\dim T_c^0\Omega = 0$, whereas for $c \in \Lambda M$, $c \neq \mathrm{const}$, $T_c^0 M$ is given by the parallel periodic vector fields ξ, $\xi(0) = \xi(1)$ and $\nabla\xi = 0$. Thus, $\operatorname{nullity} c$ is equal to the dimension of the eigenspace for the eigenvalue $+1$ under the linear mapping

$$\overset{1}{\underset{0}{\|}} c : T_0^\perp c \to T_1^\perp c = T_0^\perp c,$$

given by parallel translation of the subspace $T_0^\perp$ of $T_{c(0)}M$ into itself.

(ii) $M = S_\rho^n$ = n-dimensional sphere of radius ρ in $\mathbf{R}^{n+1}$. Then

$$R(\xi, \dot{c})\dot{c}(t) = \rho^{-2}(|\dot{c}(0)|^2\xi(t) - \langle \xi(t), \dot{c}(t)\rangle \dot{c}(t)).$$

A vector field $\xi(t)$ along a nonconstant geodesic $c(t)$ splits into a sum $\xi^{\mathsf{T}}(t) + \xi^\perp(t)$, where $\xi^{\mathsf{T}}(t) = x(t)\dot{c}(t)$ is tangential to $\dot{c}(t)$ and $\xi^\perp(t)$ is orthogonal to $\dot{c}(t)$:

$$\xi^\perp(t) = \xi(t) - \langle \dot{c}(t), \xi(t)\rangle \dot{c}(t)/|\dot{c}(t)|^2.$$

Writing $\eta(t)$ for the orthogonal component, the eigenvalue equation 3.5 splits into the two equations

$$(\top) \qquad \ddot{x}(t) + \lambda(1-\lambda)^{-1}x(t) = 0$$

and

$$(\perp) \qquad \nabla^2\eta(t) + \frac{|\dot{c}(0)|^2/\rho^2 + \lambda}{1-\lambda}\eta(t) = 0.$$

(iia) Let c be a critical point of $E \mid \Omega S^n$, i.e., a geodesic from $p \in S^n$ to $q \in S^n$. $|\dot{c}| = L(c) =$ length of c. $(\top)$ has no solutions $x \neq 0$ with $x(0) = x(1) = 0$. Nonzero solutions η of $(\perp)$ with $\eta(0) = \eta(1) = 0$ occur precisely for

$$\left(|\dot{c}|^2/\rho^2 + \lambda\right)/\left(1-\lambda\right) = \pi^2 p^2, \qquad p = 1, 2, \ldots,$$

i.e. $\lambda = \lambda_p = (\pi^2p^2 - |\dot{c}|^2/\rho^2)/(\pi^2p^2 + 1)$. The corresponding eigenvectors are $\eta(t) = \eta_p(t) = A(t)\sin \pi pt$ with $A(t)$ a parallel vector field orthogonal to $\dot{c}(t)$. Thus, the eigenspace belonging to the eigenvalue λ_p has dimension $(n-1)$. $\lambda_p < 0$ is equivalent to $|\dot{c}| < \rho\pi p$, where $\rho\pi =$ length of half a great circle on S^n_ρ. We see that index $c = k(n-1)$, with k the integer determined by

$$\pi\rho k < L(c) = |\dot{c}| \leqslant \pi\rho(k+1).$$

Nullity $c = n - 1$ if $|\dot{c}| = L(c) = \pi\rho k$, k an integer > 0, and $= 0$ otherwise.

(iib) Now let $c \neq$ const be a critical point of $E \mid \Lambda S^n_\rho$. That is, c is a closed geodesic which means it is a q-fold covered great circle. $|\dot{c}| = L(c) = 2\pi q\rho$. We see that $(\top)$ has no periodic solutions if $\lambda < 0$. For $\lambda = 0$ we get a 1-dimensional space of periodic solutions, i.e., $x(t) = x(0) =$ const. Nonzero periodic solutions of $(\perp)$ can occur only if

$$\left(|\dot{c}|^2/\rho^2 + \lambda\right)/\left(1-\lambda\right) = \left(4\pi^2q^2 + \lambda\right)/\left(1-\lambda\right) = 4\pi^2p^2, \qquad p = 0, 1, \ldots,$$

i.e., $\lambda = \lambda_p = 4\pi^2(p^2 - q^2)/(4\pi^2p^2 + 1)$. An eigenvector for the eigenvalue λ_p is of the form

$$\eta(t) \equiv \eta_p(t) = A(t)\cos 2\pi pt + B(t)\sin 2\pi pt.$$

Here, $A(t)$ and $B(t)$ are parallel vector fields along $c(t)$ orthogonal to $\dot{c}(t)$. We see that if $p = 0$, the eigenspace for λ_p has dimension $n-1$, whereas for $p > 0$ it has dimension $2(n-1)$. $\lambda_p < 0$ occurs only for $0 \leqslant p < q$. Hence, index $c = (n-1) + (q-1)2(n-1) = (2q-1)(n-1)$; nullity $c = 2(n-1)$. □

Let $c: [0, a] \to M$ be a geodesic, $\dot{c} \neq 0$, $\dim M = n$. The multiplicity k of a conjugate point $c(t_1)$ along $c \mid [0, t_1]$ is the dimension of the space of Jacobi fields $Y(t)$ along $c(t)$ vanishing at $t = 0$ and $t = t_1$. Here we allow $k = 0$, in which case $c(t_1)$ is usually called nonconjugate. For the Jacobi field $Y(t)$ we can assume $\langle Y(t), \dot{c}(t)\rangle = 0$. Clearly, $k \leqslant n - 1 = \dim \mathcal{J}_c^{0\perp}$, where $\mathcal{J}_c^{0\perp}$ denotes the space of Jacobi fields vanishing at 0 orthogonal to $\dot{c}$.

For a later application we need the

3.8 PROPOSITION. *Let $c: [0, a] \to M$ be a geodesic, $\dot{c} \neq 0$. Let $c(t_1)$ be a conjugate point of multiplicity k, $0 \leqslant k \leqslant n-1$.*

(i) *For every sufficiently small neighborhood $I(t_1)$ of t_1 on $[0, a]$, all $c(t)$ with $t \in I(t_1)$ are nonconjugate.*

(ii) *Let Y be a Jacobi field with $Y(0) = Y(t_1) = 0$. Then $\nabla Y(t_1) \in T_{t_1}^{\perp} c$ is orthogonal to every $Y'(t_1)$, $Y' \in \mathcal{J}_c^{0\perp}$.*

PROOF. Let $\dim M = n$. Consider the canonical symplectic form α on the $2n$-dimensional space $\mathcal{J}_c$ of Jacobi fields. $\mathcal{J}_c^{\perp}$ is a $(2n-2)$-dimensional nondegenerate subspace. Now, $\mathcal{J}_c^{0\perp}$ is a so-called Lagrangian subspace of $\mathcal{J}_c^{\perp}$, i.e.,

$$(*) \qquad \alpha(Y, Y') = \langle Y, \nabla Y' \rangle(t) - \langle Y', \nabla Y \rangle(t) = 0$$

for all $Y, Y' \in \mathcal{J}_c^{0\perp}$ and $\dim \mathcal{J}_c^{0\perp} = n-1$. The values $\nabla Y(t_1)$ of those $Y \in \mathcal{J}_c^{0\perp}$ which vanish at $t = t_1$ form a k-dimensional subspace of the $(n-1)$-dimensional space $T_{t_1}^{\perp} c$ of $T_{c(t_1)}M$. From $(*)$ we get $\langle \nabla Y(t_1), Y'(t_1) \rangle = 0$, if $Y(t_1) = 0$, i.e., we have (ii).

To prove (i) we choose a basis Y_i, $1 \leqslant i \leqslant n-1$, for $\mathcal{J}_c^{0\perp}$. Then the kth derivative of the determinant $| Y_1(t), \ldots, Y_{n-1}(t) |$ at $t = t_1$ is not zero. □

We can now prove the Index Theorem of Morse [**1**]. It relates the index of a geodesic c, considered as a critical point of $E \mid \Omega_{c(0)c(1)}M$, to the conjugate points along c. The right setting for conjugate points is the geodesic flow on the tangent bundle of M. Thus, the following theorem relates properties of critical points of $E \mid \Omega M$ to properties of the geodesic flow.

3.9 THEOREM. *Let $c = \{c(t), 0 \leqslant t \leqslant 1\}$ be a critical point of $\Omega = \Omega_{pq}M$. That is, c is a geodesic on M from $c(0) = p$ to $c(1) = q$. Then*

$$\begin{aligned} \text{index } c &= \sum_{0<t<1} \{\textit{multiplicity of the conjugate point } c(t)\} \\ &= \textit{sum of proper conjugate points } c(t_1), 0 < t_1 < 1, \\ &\qquad \textit{each counted with its multiplicity} \geqslant 1. \end{aligned}$$

PROOF. $c(t_1)$ is conjugate $c(0)$ along $c \mid [0, t_1]$ of multiplicity k if the space $W(t_1)$ of Jacobi fields $Y(t)$ along $c(t)$ with $Y(0) = Y(t_1) = 0$ has dimension k. Here, we may restrict ourselves to the Jacobi fields $Y(t) \perp \dot{c}(t)$.

Define a mapping $\Phi : W(t_1) \to T_c\Omega$; $Y \mapsto \xi_Y$ by

$$\xi_Y(t) \equiv \xi(t) = \begin{cases} Y(t), & 0 \leqslant t \leqslant t_1, \\ 0, & t_1 \leqslant t \leqslant 1. \end{cases}$$

ξ_Y is a broken Jacobi field with $\nabla \xi_Y(t_1 -) - \nabla \xi_Y(t_1 +) = \nabla Y(t_1) \neq 0$ unless $Y = 0$. Clearly, Φ is linear and injective. Moreover, when $t_1 \neq t_1'$, $\Phi W(t_1) \cap \Phi W(t_1') = 0$. Denote by U the subspace of $T_c\Omega$ generated by the $\Phi W(t)$, $0 < t < 1$. Then

$$\dim U = \sum_{0<t<1} \dim W(t) = \text{right-hand side of 3.9}.$$

We claim that $\dim U \leqslant \text{index } c$. We first show that $D^2E(c) \mid U \equiv 0$. To see this consider $\xi \in \Phi W(t_1)$, $\xi' \in \Phi W(t_1')$ with $t_1' \leqslant t_1$. Then

$$D^2E(c)(\xi, \xi') = -\int_0^1 \langle \nabla^2 \xi + R(\xi, \dot{c})\dot{c}, \xi' \rangle(t)\, dt + \langle \nabla \xi, \xi' \rangle(t) \Big|_0^{t_1-} = 0,$$

since $\xi'(t_1) = \xi'(0) = 0$. Moreover, $U \cap T_c^0 = 0$ since the elements of T_c^0 are the Jacobi fields $Y(t)$ with $Y(0) = Y(1) = 0$ (cf. 3.6). Our claim therefore follows from Remark 2 after 3.4.

We conclude the proof by showing that if there is a $\xi^* \in T_c^-$ with $D^2E(c)(\xi^*, U) = 0$, then $\xi^* = 0$. That is, $\dim U = \dim \mathrm{pr}_- U = \dim T_c^- = \operatorname{index} c$, with $\mathrm{pr}_- : T_c \to T_c^-$ the orthogonal projection. To prove $\xi^* = 0$, we begin by showing that $\xi^*(t)$ can be written as $\xi^*(t) = \Sigma_i w^i(t) Y_i(t)$, where the Y_i, $1 \leqslant i \leqslant n-1$, form a basis for $\mathcal{J}_c^{0\perp}$. First $\xi^* \in T_c^-$ means that $\xi^*(t) \perp \dot c(t)$. Next, consider the case when the $\{Y_i(t_1); 1 \leqslant i \leqslant n-1\}$ are linearly dependent, i.e., $c(t_1)$ is a conjugate point of multiplicity $k > 0$. For $Y \in \mathcal{J}_c^{0\perp}$ with $Y(t_1) = 0$, consider $\xi \in \Phi W(t_1) \subset U$ with $\xi \,|\, [0, t_1] = Y \,|\, [0, t_1]$, $\xi \,|\, [t_1, 1] = 0$. Then

$$0 = D^2E(c)(\xi, \xi^*) = \langle \nabla \xi, \xi^* \rangle \big|_0^{t_1^-} = \langle \nabla Y(t_1), \xi^*(t_1) \rangle = 0.$$

That is, according to 3.8, $\xi^*(t_1)$ belongs to the space spanned by the $\{Y_i(t_1), 1 \leqslant i \leqslant n-1\}$. We conclude by computing $D^2E(c)(\ ,\ )$. As an element of T_c^-, ξ^* certainly is differentiable. Thus, using partial integration,

$$\begin{aligned}
0 \geqslant D^2E(c)(\xi^*, \xi^*) &= -\int_0^1 \langle \nabla^2 \xi^* + R(\xi^*, \dot c)\dot c, \xi^* \rangle(t)\, dt \\
&= -\int_0^1 \Big\langle \sum_i w^i(\nabla^2 Y_i + R(Y_i, \dot c)\dot c), \sum_j w^j Y_j \Big\rangle(t)\, dt \\
&\quad - \int_0^1 \sum_{i,j} \dot w^i w^j \langle \nabla Y_i, Y_j \rangle(t)\, dt - \int_0^1 \Big\langle \frac{d}{dt}\Big(\sum_i \dot w^i Y_i\Big), \sum_j w^j Y_j \Big\rangle(t)\, dt \\
&= \int_0^1 \sum_{i,j} \dot w^i w^j (\langle Y_i, \nabla Y_j \rangle - \langle Y_j, \nabla Y_i \rangle)(t)\, dt + \int_0^1 \Big\langle \sum_i \dot w^i Y_i, \sum_j \dot w^j Y_j \Big\rangle(t)\, dt \\
&\geqslant 0.
\end{aligned}$$

The second to last term vanishes, since $\mathcal{J}_c^{0\perp}$ is a Lagrangian subspace, i.e., a subspace of $\mathcal{J}_c^{\perp}$ on which the form α vanishes identically. Hence, $D^2E(c)(\xi^*, \xi^*) = 0$, i.e., $\xi^* = 0$. □

EXAMPLE. We determine the index of a geodesic segment $c = \{c(t), 0 \leqslant t \leqslant 1\}$ of length $L(c) = l$ on S_ρ^n. The Jacobi equation reads (cf. 3.7)

$$(*) \qquad \nabla^2 Y(t) + (l^2/\rho^2) Y(t) = 0.$$

Here we have considered only Jacobi fields orthogonal to $\dot c(t)$. We see that they are the only ones which can vanish at $t = 0$ and at some $t_1 > 0$, unless we have the 0-field.

The solutions of $(*)$ are of the form $Y(t) = A(t)\sin(lt/\rho)$ with $A(t)$ a parallel vector field. If $A(t) \neq 0$, $Y(t) = 0$ can occur only for $t_p = \pi\rho p/l$, p an integer. Each proper conjugate point has multiplicity $n-1$. Thus, index $c = k(n-1)$, where k is the integer determined by $\pi\rho k < l \leqslant \pi\rho(k+1)$. Compare this with 3.7(iia). □

As a first consequence of Morse's Index Theorem we prove

3.10 THEOREM. *Let $c: I \to M$ be a geodesic from $c(0) = p$ to $c(1) = q$. We also consider c as a critical point of $E \mid \Omega_{pq}M$.*

(i) *If there are no conjugate points in the interior or on the boundary of I then there exists a neighborhood $\mathfrak{U}(c)$ of c in $\Omega_{pq}M$ such that for all $b \in \mathfrak{U}(c)$, $E(b) \geqslant E(c)$, with $E(b) = E(c)$ only for $b = c$.*

(ii) *Let $k > 0$ be the number of conjugate points in the interior of I, each counted with its multiplicity. Then there exists an immersion $F: (B^k, 0) \to (\Omega_{pq}M, c)$ of the unit ball $B^k = \{x \in \mathbf{R}^k, |x| < 1\}$ such that $E(F(x)) \leqslant E(c)$ and, hence, $L(F(x)) \leqslant L(c)$ for all $x \in B^k$, with equality only for $x = 0$.*

PROOF. Under the hypothesis in (i) we find from 3.8 that $D^2E(c)$ is positive definite. Hence, if we choose a sufficiently small natural chart $(\exp_c^{-1}, \mathfrak{U}(c))$ around c, the Taylor series for $E(b)$, $b \in \mathfrak{U}(c)$, $\xi = \exp_c^{-1}b$, starts with

$$E(b) = E(c) + \tfrac{1}{2}D^2E(c)(\xi, \xi) + o\|\xi\|_1^2.$$

This is $\geqslant E(c)$ for all ξ, $\|\xi\|_1$ sufficiently small.

The hypothesis in (ii) implies, according to 3.8, that index $c = k$. Let $\{\xi_i, 1 \leqslant i \leqslant k\}$ be an orthonormal basis of T_c^- where ξ_i is an eigenvector for the eigenvalue $-\lambda_i < 0$. For $\delta > 0$ sufficiently small

$$F: x = (x^1, \ldots, x^k) \in B^k \mapsto \exp_c\Big(\delta \sum_i x^i \xi_i\Big) \in \Omega_{pq}M$$

will be an immersion. From the Taylor expansion

$$E(F(x)) = E(c) - \frac{1}{2}\delta^2 \sum \lambda_i (x^i)^2 + o\|\delta\xi\|_1^2,$$

we have proved our claim, provided δ is sufficiently small. $\square$

We continue with an index theorem for a closed geodesic. First we observe that a closed geodesic c can also be viewed as a critical point of $E \mid \Omega$, $\Omega = \Omega_{c(0)c(0)}M$. Recall from 3.4 that we denote by $\operatorname{index}_\Omega c$ the index of c as a critical point of $E \mid \Omega$, whereas for index c we also write $\operatorname{index}_\Lambda c$.

3.11 PROPOSITION. *Let c be a closed geodesic. Then* $\operatorname{index}_\Omega c \leqslant \operatorname{index}_\Lambda c \equiv$ index c.

PROOF. $D^2E(c)$ has the same form, whether we derive it from $E \mid \Omega M$ or $E \mid \Lambda M$. The only difference is that in the first case we restrict it to $T_c\Omega M$, the space of H^1-vector fields $\xi(t)$ along $c(t)$ which vanish for $t = 0$ and $t = 1$, whereas in the second case we allow $\xi(t)$ with $\xi(0) = \xi(1)$.

Now, as we remarked after 3.4, since $T_c^-\Omega$ is a subspace of $T_c\Lambda$ on which $D^2E(c)$ is negative definite, $\operatorname{index}_\Lambda c \geqslant \dim T_c^-\Omega =$ index c. $\square$

Let c be a closed geodesic. We want to write a formula for the term which has to be added to $\operatorname{index}_\Omega c$ to obtain $\operatorname{index}_\Lambda c$. It was introduced by Morse under the name concavity. As a preliminary step we introduce the following concepts. $\mathfrak{J}_c^\perp$ denotes the $(2n-2)$-dimensional space of Jacobi fields along c which are orthogonal to c. $\mathcal{J}_c^{0\perp}$ denotes the $(n-1)$-dimensional subspace of those fields which vanish at $t=0$.

3.12 DEFINITION. Let $c(t)$, $0 \leqslant t \leqslant 1$, be a closed geodesic. Denote by $\mathcal{J}_c^{\perp}$ the Jacobi fields $Y(t)$ along $c(t)$ with $\langle \dot c(t), Y(t)\rangle = 0$. Put $\mathcal{J}_c^{\perp} \cap T_c\Lambda = \mathcal{J}_{c,\Lambda}^{\perp}$, i.e. $\mathcal{J}_{c,\Lambda}^{\perp}$ consists of the Jacobi fields $Y(t)$ along $c(t)$ with $Y(0) = Y(1)$, $\mathcal{J}_{c,\Lambda}^{\perp}$ contains the space $\mathcal{J}_{c,\mathrm{per}}^{\perp}$ of periodic Jacobi fields $Y(t)$ along $c(t)$, orthogonal to $\dot c(t)$. $\dim \mathcal{J}_{c,\mathrm{per}}^{\perp} = \operatorname{null} c$ (cf. 3.6). Define the concavity of c, concav c, by

$$\operatorname{concav} c + \operatorname{null} c = \dim\bigl(\text{null space plus negative eigenspace of } D^2E(c) \mid \mathcal{J}_{c,\Lambda}\bigr).$$

REMARK. Put $c(0) = c(1) = p$. Consider the subspace $T_0^{\perp}c = T_1^{\perp}c$ of T_pM formed by the vectors orthogonal to $\dot c(0) = \dot c(1)$. $\dim T_0^{\perp}c = n-1$ if $\dim M = n$. Write $V_h \oplus V_v$ for the direct sum of two copies of $T_0^{\perp}c$. Then we have the isomorphism

$$Y \in \mathcal{J}_c^{0\perp} \mapsto \tilde Y(0) \equiv (Y(0), \nabla Y(0)) \in V_h \oplus V_v.$$

On $\mathcal{J}_c^{\perp} \cong V_h \oplus V_v$ we have the symplectic form α defined by

$$\alpha(Y,Z) \equiv \alpha(\tilde Y(0), \tilde Z(0)) = \langle Y(t), \nabla Z(t)\rangle - \langle Z(t), \nabla Y(t)\rangle,$$

where the right-hand side is independent of $t \in S$. We consider the transformation

$$P : V_h \oplus V_v \to V_h \oplus V_v; \quad (Y(0), \nabla Y(0)) \mapsto (Y(1), \nabla Y(1)).$$

Then $\alpha(P\tilde Y(0), P\tilde Z(0)) = \alpha(\tilde Y(0), \tilde Z(0))$, i.e., P is a so-called symplectic transformation. With this we can write

$$\mathcal{J}_{c,\mathrm{per}} = \ker(P - \mathrm{id}), \quad \mathcal{J}_{c,\Lambda}^{\perp} = \ker \mathrm{pr}_h \circ (P - \mathrm{id}),$$

where $\mathrm{pr}_h : V_h \oplus V_v \to V_h$ is the projection.

3.13 PROPOSITION. *With the previously introduced notation we define subspaces T_1 and T_2 of $T_0^{\perp}c$ by*

$$T_1 = \mathrm{pr}_h \circ \ker(P - \mathrm{id}); \quad T_2 = V_v \cap \operatorname{im}(P - \mathrm{id}).$$

Then $T_0^{\perp}$ is the orthogonal sum of T_1 and T_2.

Note. Here we have used the canonical identification of V_h and V_v with $T_0^{\perp}c$.

PROOF. We claim that $\ker(P-\mathrm{id})$ and $\operatorname{im}(P-\mathrm{id})$ are the α-orthogonal subspaces of each other. Indeed, their dimensions add up to $2n-2 = \dim(V_h \oplus V_v)$. Moreover, if $\tilde Y(0) \in \ker(P-\mathrm{id})$ and $\tilde W(0) = P\tilde Z(0) - \tilde Z(0) \in \operatorname{im}(P-\mathrm{id})$,

$$\begin{aligned}\alpha(\tilde Y(0), \tilde W(0)) &= \langle Y(0), \nabla Z(1) - \nabla Z(0)\rangle - \langle Z(1) - Z(0), \nabla Y(0)\rangle \\ &= \{\langle Y(1), \nabla Z(1)\rangle - \langle Z(1), \nabla Y(1)\rangle\} \\ &\quad - \{\langle Y(0), \nabla Z(0)\rangle - \langle Z(0), \nabla Y(0)\rangle\} \\ &= 0.\end{aligned}$$

Hence, if $Y(0) = Y(1) \in T_1$ and $\nabla Z(1) - \nabla Z(0) \in T_2$, we get $\langle Y(0), \nabla Z(1) - \nabla Z(0)\rangle = 0$, i.e., $\langle T_1, T_2\rangle \doteq 0$. Similarly, with $T_1' = V_v \cap \ker(P - \mathrm{id})$ and $T_2' = \mathrm{pr}_h \circ \mathrm{im}(P - \mathrm{id})$, we get $\langle T_1', T_2'\rangle = 0$. Thus, $\dim T_1 + \dim T_2 \leq n - 1$, $\dim T_1' + \dim T_2' \leq n - 1$. On the other hand, $\dim T_1 + \dim T_1' = \dim \ker(P - \mathrm{id})$, $\dim T_2 + \dim T_2' = \dim \mathrm{im}(P - \mathrm{id})$ and, therefore, $\dim T_1 + \dim T_2 = n - 1$. □

We now can prove the Index Theorem for a closed geodesic. In the case that the closed geodesic c is nondegenerate, it is due to Morse [**1**]. For the case of an arbitrary geodesic, see Morse [**2**] and Klingmann [**1**]. An entirely different version, where the relation between $\mathrm{index}_\Lambda c$ and $\mathrm{index}_\Omega c$ plays no role, is given in Klingenberg [**2**]: The index is interpreted as the intersection number of a closed curve in the Lagrangian-Grassmann manifold SU/SO with its codimension one submanifold, representing the 1-cohomology class of SU/SO.

3.14 THEOREM. *Let c be a closed geodesic. Then* $\mathrm{index}_\Lambda c = \mathrm{index}_\Omega c + \mathrm{concav}\, c \leq \mathrm{index}_\Omega c + n - 1$.

PROOF. We begin as in the proof of 3.9 with the linear mappings $\Phi : W(t_1) \to T_c\Omega M$, $0 < t_1 < 1$. Denote by U the space generated by the $\Phi W(t)$, $0 < t < 1$. $\dim U = \mathrm{index}_\Omega c$. Let U' be a complement of the space $\mathfrak{J}^\perp_{c,\mathrm{per}}$ in the null space plus negative eigenspace of $D^2E(c) \mid \mathfrak{J}^\perp_{c,\Lambda}$. Clearly, $U \cap U' = 0$. Hence, $\dim U + U' = \mathrm{index}_\Omega c + \mathrm{concav}\, c$. Moreover, $D^2E(c) \mid U + U' \leq 0$. Indeed, we already know that $D^2E(c) \mid U \equiv 0$. If $\xi \in U$, and $\eta \in \mathfrak{J}^\perp_{c,\Lambda}$, then $D^2E(c)(\xi, \eta) = \langle \nabla\eta(1) - \nabla\eta(0), \xi(0)\rangle = 0$, since $\xi(0) = 0$. From the definitions it also follows that, under $\mathrm{pr}_- : T_c\Lambda \to T_c^-\Lambda$, $U + U'$ is mapped injectively.

To conclude the proof, we therefore need only show that if $\xi^* \in T_c^-\Lambda$, $D^2E(c)(\xi^*, U + U') = 0$, then $\xi^* = 0$. To see this let $\eta(t) = Z(t) \in \mathfrak{J}^\perp_{c,\Lambda}$. With $\eta^- = \mathrm{pr}^-(\eta) \in T_c^-\Lambda$ we have

$$0 = D^2E(c)(\eta, \xi^*) = \langle \nabla Z(1) - \nabla Z(0), \xi^*(0)\rangle = 0.$$

That is, $\xi^*(0)$ belongs to the subspace T_1 of $T_0^\perp c$, introduced in 3.13. Hence, there exists a $Y \in \mathfrak{J}^\perp_{c,\mathrm{per}}$ with $\xi^*(0) = Y(0)$, i.e., $\zeta = \xi^* - Y \in T_c\Omega$. Since $Y \in$ nullspace of $D^2E(c)$,

$$0 \geq D^2E(c)(\xi^*, \xi^*) = D^2E(c)(\zeta, \zeta).$$

$\mathrm{pr}^-(\zeta) \in T_c^-\Lambda$ coincides with $\mathrm{pr}^-(\xi^*) = \xi^*$. Moreover, $\mathrm{pr}^-(\zeta) \in \mathrm{pr}^-(U)$. But by hypothesis, $D^2E(c)(\xi^*, \mathrm{pr}^-(U)) = 0$. Therefore, $\xi^* = 0$.

Finally, we observe that an upper bound for $\mathrm{concav}\, c$ is given by the dimension of $T_1 \subset T_0^\perp c$. □

We conclude with a certain generalization of 2.16. There we proved essentially the following: Let κ_0, κ_1 be real numbers, $0 \leq \kappa_0 \leq \kappa_1$, such that there are no critical points c with $\kappa_0 \leq E(c) \leq \kappa_1$. Then the $-\mathrm{grad}\, E$ deformation ϕ_s, $s \geq 0$, deforms $\Lambda^{\kappa_1}M$ into $\Lambda^{\kappa_0 -}M$. In particular, all relative homotopy groups vanish,

$$\pi_k = \pi_k(\Lambda^{\kappa_1}M, \Lambda^{\kappa_0 -}M) = 0, \qquad k = 0, 1 \ldots.$$

The elements of π_k are the classes of homotopic maps $f:(I^k, \partial I^k) \to (\Lambda^{\kappa_1}M, \Lambda^{\kappa_0 -}M)$. In our generalization we do not exclude altogether the existence of critical points c with $\kappa_0 \leqslant E(c) \leqslant \kappa_1$; we only forbid their index from being too small.

3.15 THEOREM. *Let κ_0, κ_1 be real numbers, $0 \leqslant \kappa_0 < \kappa_1$ and l an integer > 0. Assume that there are no critical points c of index $< l$ with $\kappa_0 \leqslant E(c) \leqslant \kappa_1$ and no critical points at all with E-value κ_1. Then*

$$(*) \qquad f:(I^k, \partial I^k) \to (\Lambda^{\kappa_1}M, \Lambda^{\kappa_0 -}M), \qquad k < l,$$

is homotopic to a mapping having its image entirely in $\Lambda^{\kappa_0 -}M$, i.e., $\pi_k(\Lambda^{\kappa_1}M, \Lambda^{\kappa_0 -}M) = 0$, for $k < l$.

Notes. 1. A standard result of algebraic topology (cf. Spanier [**1**]), says: If $\pi_k(\Lambda^{\kappa_1}, \Lambda^{\kappa_0 -}) = 0$ for $0 \leqslant k < l$ and $\pi_1(\Lambda^{\kappa_0 -}) = 0$, then also $H_k(\Lambda^{\kappa_1}, \Lambda^{\kappa_0 -}) = 0$ for $0 \leqslant k < l$ and $\pi_l(\Lambda^{\kappa_1}, \Lambda^{\kappa_0 -}) = H_l(\Lambda^{\kappa_1}, \Lambda^{\kappa_0 -})$. Thus, in 3.15 we actually also prove the vanishing of certain homology groups.

2. In 2.18 we showed that in a ϕ-family $\mathcal{Q}$ there exist members A which can be pushed under the ϕ_s-deformation into $\mathcal{U} \cap \Lambda^{\kappa -}$, where $\mathcal{U}$ is a neighborhood of the set $\operatorname{Cr} \kappa$ of critical points of E-value $\kappa = \kappa_{\mathcal{Q}}$. It should be possible to prove a similar refinement of 3.15. More precisely, our methods employed for the proof of 3.15 should suffice to show the following: Let κ_0, κ_1 be such that there are no critical points c of index $< l$ with E-value $\kappa_0 < E(c) \leqslant \kappa_1$ and no critical point at all with E-value κ_1. Then, given any neighborhood $\mathcal{U}$ of the set $\operatorname{Cr} \kappa_0$ of critical points of index $< l$ at E-value κ_0, it is possible to deform the mapping $(*)$ such that its image belongs to $\mathcal{U} \cap \Lambda^{\kappa_0 -}$.

PROOF. Let $\kappa \in]\kappa_0, \kappa_1[$ be a critical value. All we need show is: For a certain open neighborhood $\mathcal{V}$ of the critical set $\operatorname{Cr} \kappa$ at $\{E = \kappa\}$ in $\Lambda^{\kappa_1}M$ and a mapping

$$(\dagger) \qquad f_0:(I^k, \partial I^k) \to (\mathcal{V} \cup \Lambda^{\kappa -}, \Lambda^{\kappa_0 -}),$$

there exists a homotopy

$$(\dagger\dagger) \qquad f_s:(I^k, \partial I^k) \to (\Lambda^{\kappa_1}, \Lambda^{\kappa_0 -}), \qquad 0 \leqslant s \leqslant 1,$$

of f_0 with $\operatorname{im} f_1 \subset \Lambda^{\kappa -}$. This homotopy in general will be nonstandard, i.e., it will not be up to parameterization, of the form $\phi_s \circ f_0$. Once we have proved this we can complete the proof of the theorem as follows: Consider for a mapping $(*)$ the ϕ-family $\{\phi_s \circ f(I^k); s \geqslant 0\}$. Let κ be its critical value. Then $\kappa < \kappa_1$. If $\kappa < \kappa_0$, we are done. Otherwise, for any prescribed neighborhood $\mathcal{U}$ of $\operatorname{Cr} \kappa$, $\phi_s \circ f$ will be of the form $(\dagger)$ for all $s \geqslant$ some s_0 (cf. 2.18). We write f_0 instead of $\phi_s \circ f$ for some $s \geqslant s_0$. Now apply the homotopy $(\dagger\dagger)$ which shows that our original f (see $(*)$) is homomotopic to an f' with $\max E \,|\, f' < \kappa$.

To show the existence of the homotopy $(\dagger\dagger)$ consider a $c \in \operatorname{Cr} \kappa$. Let $B_\rho(c)$ be a strongly convex neighborhood of c and denote by u the normal coordinates for

$B_\rho(c)$ based at c. We write $u = (u^-, u^0, u^+)$, in correspondence with the decomposition $T_c\Lambda = T_c^- \oplus T_c^0 \oplus T_c^\perp$ of the tangent space at c (cf. 3.3). In these coordinates $E \mid B_\rho(c)$ can be written as

$$E(u) = E(u^-, u^0, u^+) = \kappa - \frac{1}{2}\sum_i \lambda_i (u_i^-)^2 + \frac{1}{2}\sum_k \lambda_k (u_k^+)^2 + E_{(3)}(u),$$

with $E_{(3)}(u) = o(\|u\|_1^2)$. The $-\lambda_i$, $1 \leqslant i \leqslant m = \operatorname{index} c$, are the negative eigenvalues of $D^2E(c)$, while the λ_k, $k \geqslant m + 2 + \operatorname{null} c$, are the positive eigenvalues of $D^2E(c)$.

We introduce new coordinates $v = (v^-, v^0, v^+)$ by $v^- = u^- - h(u^0, u^+)$, $v^0 = u^0$, $v^+ = u^+$, with $h(0,0) = 0$, $\partial E(h(u^0, u^+), u^0, u^+)/\partial u_i^- = 0$, $1 \leqslant i \leqslant m$. Then $\tilde{E}(v) = E(u(v))$ is of the form

$$\tilde{E}(v) = \tilde{E}(v^-, v^0, v^+) = \kappa - \frac{1}{2}\sum_i \lambda_i (v_i^-)^2 + \frac{1}{2}\sum_k \lambda_k (v_k^+)^2 + \tilde{E}_{(3)}(v)$$

with $\tilde{E}_{(3)}(v) = o(\|v\|_1^2)$ and $\partial \tilde{E}_{(3)}(0, v^0, v^+)/\partial v_i^- = 0$, $1 \leqslant i \leqslant m$. That is, $\tilde{E}_{(3)}(v)$ contains no terms linear in v^-.

The existence of the implicitly defined functions $h_i(u^0, u^+)$ follows from the fact that the $(m \times m)$-matrix with elements $\partial^2 E(0,0,0)/\partial u_i^- \partial u_j^- = -\lambda_i \delta_{ij}$ is invertible.

We can restrict the domain of the new coordinates v such that the matrix $(\partial^2 \tilde{E}(v)/\partial v_i^- \partial v_j^-)$ is negative definite for all v. Moreover, we have a $\sigma > 0$ and an $\alpha = \alpha(c, \sigma)$ such that $\|v^-\|_1 > \alpha\|(v^0, v^+)\|_1$ implies $\tilde{E}(v) \leqslant \kappa$ whenever $\|v^-\|_1 \leqslant \sigma$. We can make α unique by choosing it minimal with this property. Thus, $\tilde{E}(v) < \kappa$ on the open cone $\{\|v^-\|_1 > \alpha\|(v^0, v^+)\|_1\}$.

We now define the closed neighborhood $C_\sigma(c)$ of c by $\|v^-\|_1 \leqslant \sigma$, $\|(v^0, v^+)\|_1 \leqslant \sigma/2\alpha$. For each $v = (v^-, v^0, v^+)$ in the range of $C_\sigma(c)$ with $v^- \neq 0$, the curve $v(t) = (tv^-, v^0, v^+)$, $1 \leqslant t \leqslant \sigma/\|v^-\|_1 = t_1$, from $v = v(1)$ to $v(t_1)$ lies inside $C_\sigma(c)$. $\|v^-(t_1)\|_1 > \alpha\|(v^0(t_1), v^+(1))\|_1$, hence, $\tilde{E}(v(t_1)) < \kappa$. Moreover, $d\tilde{E}(v(t))/dt < 0$, since $\partial \tilde{E}(0, v^0, v^+)/\partial v_i^- = 0$ and $(\partial^2 \tilde{E}(v)/\partial v_i^- \partial v_j^-)$ is negative definite.

On $C_\sigma(c)$ we define the vector field X as the field which in the v-coordinates has the representation $X(v) = (v^-, 0, 0)$. Thus, whenever $e \in C_\sigma(c)$ has a v-coordinate with $v^-(e) \neq 0$, $D\tilde{E}.X(e) < 0$, $D^2\tilde{E}.(X, X) < 0$. Hence, the flow along a nonconstant integral curve of X decreases the energy until the energy becomes $< \kappa$.

Denote by $\frac{1}{2}C_\sigma(c)$ the subset $\{\|v^-\|_1 \leqslant \sigma/2;\ \|(v^0, v^+)\|_1 \leqslant \sigma/4\alpha\}$ of $C_\sigma(c)$. Since $\operatorname{Cr}\kappa$ is compact, a finite subset of the open covering $\{\operatorname{int}\frac{1}{2}C_\sigma(c);\ c \in \operatorname{Cr}\kappa\}$ will suffice to cover $\operatorname{Cr}\kappa$. We enumerate this set by $\{\operatorname{int}\frac{1}{2}C_{\sigma_r}(c_r),\ r = 1, \ldots, R\}$. We also write briefly C_r, $\frac{1}{2}C_r$ instead of $C_{\sigma_r}(c_r)$, $\frac{1}{2}C_{\sigma_r}(c_r)$.

Let $\gamma_r\colon \Lambda M \to \mathbf{R}$ be a differentiable function with values in $[0, 1]$ such that $\gamma_r \mid \frac{1}{2}C_r = 1$, $\gamma_r \mid \partial C_r = 0$. For each r we define the vector field Y_r on ΛM by $Y_r(e) = \gamma_r(e)X_r(e)$ if $e \in C_r$, and by $Y_r(e) = 0$ if $e \in \complement C_r$. Here, X_r is the vector field defined above on C_r.

Denote by $\mathcal{U}$ the union of the int $\frac{1}{2}C_r$, $r = 1,\ldots,R$. There exists an open neighborhood $\mathcal{V}$ of Cr κ and a $\rho > 0$ such that $d_\Lambda(c', c'') \geqslant \rho$ if $c' \in \mathcal{V}$, $c'' \in \complement\mathcal{U}$. Indeed, the distance between a point $c'' \in \complement\mathcal{U}$ and Cr κ is $\geqslant$ some positive constant independent of the choice of c''. Now take for $\mathcal{V}$, e.g., the points having distance $< \frac{1}{2}$ this constant. It follows (cf. the proof of 2.18) that there exists an $\varepsilon > 0$ such that the trajectory $\phi_s c''$, $s \geqslant 0$, of a $c'' \in \complement\mathcal{U}$ with $E(c'') < \kappa + \varepsilon$ under the ($-$grad E)-flow has E-value $< \kappa$ when it enters the neighborhood $\mathcal{V}$.

Now consider an f_0 as in (†). By applying, if necessary, a deformation ϕ_s we can even assume max $E|f_0(I^k) \leqslant \kappa + \varepsilon$. Beginning with $r = 1$, we define a deformation ${}_1\psi_s \circ f_0$, $s \geqslant 0$, of f_0 by the integral flow ${}_1\psi_s : \Lambda M \to \Lambda M$ of the vector field Y_1. Whenever $f_0(x) \in \frac{1}{2}C_1$ and the ${}_1v$-coordinate ${}_1v(f_0(x)) = ({}_1v^-(f_0(x)), {}_1v^0(f_0(x)), {}_1v^+(f_0(x)))$ on C_1 satisfies ${}_1v^-(f_0(x)) \neq 0$, $E({}_1\psi_s f_0(x)) < \kappa$ for sufficiently large s. Observe now that the condition ${}_1v^-(f_0(x)) \neq 0$ for $x \in f_0^{-1}(\frac{1}{2}C_1)$ is a transversality condition for f_0, since dim $I^k = k <$ index $c_1 =$ codim$\{{}_1v^- = 0\}$. Hence, this condition can be met by an arbitrarily small appropriate deformation of f_0.

We thus have the existence of a deformation ${}_1f_0$ of f_0 such that $E({}_1f_0(x)) < \kappa$ whenever $f_0(x) \in$ int $\frac{1}{2}C_1$. The deformation ${}_1\psi_s$ also operates on $C_1 - \frac{1}{2}C_1$. It may have happened that thereby an $f_0(x)$ in this set is deformed into $\complement\mathcal{U}$. We will take care of this at the end. In any case, the E-value of the deformed element ${}_1f_0(x)$ can only have become smaller. In particular, this value will be $< \kappa + \varepsilon$.

Using the integral flow ${}_2\psi_s : \Lambda M \to \Lambda M$ of the vector field Y_2, we can deform ${}_1f_0$ into ${}_2\psi_s \circ {}_1f_0$ such that, for sufficiently large $s > 0$, $E({}_2\psi_s \circ {}_1f_0(x)) < \kappa$, whenever ${}_1f_0(x) \in$ int $\frac{1}{2}C_2$ and the negative component ${}_2v^-({}_1f_0(x))$ of the ${}_2v$-coordinate of ${}_1f_0(x)$ is $\neq 0$. This latter condition again can be met for transversality reasons.

Proceeding in this manner, we see that our original f_0 (in (†)) with max $E|f_0(I^k) < \kappa + \varepsilon$ can be deformed into an ${}_Rf_0$ such that, for all $x \in I^k$, $E({}_Rf_0(x)) < \kappa$ or else ${}_Rf_0(x) \in \complement\mathcal{U}$, $E({}_Rf_0(x)) < \kappa + \varepsilon$. Now on ${}_Rf_0(I^k)$ apply a deformation ϕ_{s_0} which transforms the image into $\mathcal{V} \cap \Lambda^{\kappa-}$. Under this deformation any ${}_Rf_0(x) \in \complement\mathcal{U}$ will be transformed into $\Lambda^{\kappa-}$; thus with $f_1 = \phi_{s_0} \circ {}_Rf_0$, we get the desired deformation (††). □

Chapter 3. Appendix: The S^1-and the $\mathbf{Z}_2$-action on ΛM

We exhibit an additional structure of the Hilbert manifold ΛM of closed H^1-curves, i.e., the existence of a canonical $\mathbf{O}(2)$-action which preserves the Riemannian metric on ΛM and leaves the function E invariant. This action stems from the canonical $\mathbf{O}(2)$-action on the source S of the elements $\{c: S \to M\}$ in ΛM. The full importance of this structure for the problem of the existence of many—even infinitely many—geometrically distinct closed geodesics on a compact Riemannian manifold M has become apparent only recently; cf. Klingenberg [3, 5] and Chapter 5 where we will consider the Morse complex and its applications. Cf. also the remark at the end of this Appendix.

3.A.1 Definition. Define an S^1-action on ΛM by $\chi^\sim : S^1 \times \Lambda M \to \Lambda M$, $(z, c) \mapsto z.c$. Here, with $z = e^{2\pi i r} \in S^1$, $z.c$ is defined by $z.c(t) = c(t + r)$. The orbit of $c \in \Lambda M$ under this S^1-action will also be denoted by $S^1.c$.

Note. We can say that the elements of $S^1.c$ are obtained from c by 'change of the initial point'.

3.A.2 Lemma. *The action $\chi^\sim$ is continuous but not differentiable. On the other hand, for a fixed $z \in S^1$, the mapping $\chi_z^\sim : \Lambda M \to \Lambda M$, $c \mapsto z.c$ is an isometry which leaves the function $E : \Lambda M \to \mathbf{R}$ invariant.*

Note. Instead of $\chi_z^\sim$ we also write $z : \Lambda M \to \Lambda M$.

Proof. Consider the natural charts $(\exp_c^{-1}, \mathfrak{U}(c))$, $(\exp_{z.c}^{-1}, \mathfrak{U}(z.c))$ for $c \in C^\infty(S, M)$, $z \in S^1$ (cf. 1.10). Then the mapping $z : \Lambda M \to \Lambda M$ is represented by the linear isomorphism

$$(\xi(t)) \in H^1(c^*TM) \mapsto (\xi(t + r)) \in H^1((z.c)^*TM).$$

From the local representation of E and the Riemannian metric $\langle\ ,\ \rangle_1$ in Chapter 1, it now follows that $z : \Lambda M \to \Lambda M$ is an isometry with $E(z.c) = E(c)$.

It remains to show that $\chi^\sim$ is continuous. Since $\chi_z^\sim$ is an isometry we have

$$d_\Lambda(c, z.e) \leq d_\Lambda(c, z.c) + d_\Lambda(c, e); \quad d_\Lambda(e, z.e) \leq d_\Lambda(c, z.c) + 2d_\Lambda(e, c).$$

Hence, it suffices to show that $d_\Lambda(c, z.c)$ tends to zero as z tends to $1 \in S^1$ for $c \in C'^\infty(S, M)$. But since $\exp_c^{-1}(c) = (\xi(t)) = 0$, the coordinate $\exp_c^{-1}(z.c)$ is given by the horizontal vector field $r\partial t \in H^1(c^*TM)$, where $z = e^{2\pi i r}$. As r goes to 0, so does $r\partial t$.

Finally, we observe that the S^1-action $\chi^\sim$ cannot be differentiable. In that case, it would follow that the composition

$$r \in S \mapsto z = e^{2\pi i r} \in S^1 \mapsto z.e \in \Lambda M \mapsto (z.e)(0) = e(r) \in M$$

is also differentiable for every $e \in \Lambda M$. But for a nondifferentiable $e : S \to M$ this is false. □

3.A.3 COROLLARY. *If c is a critical point of $E : \Lambda M \to \mathbf{R}$, then every $z.c$ on the S^1-orbit of c is critical.*

PROOF. From 3.A.2 follows that $DE(c) = 0$ implies $DE(z.c) = 0$. □

3.A.4 COROLLARY. *The isotropy subgroup $I^\sim(c)$ of c under the S^1-action is either the whole group S^1 (this is true if and only if $c \in \Lambda^0 M$ = space of constant mappings $c : S \to M$), or $I^\sim(c)$ is a finite cyclic group.*

Note. Recall that the isotropy group of c is defined as the set of $z \in S^1$ with $z.c = c$.

PROOF. Since $\chi_c^\sim : z \in s \to z.c \in \Lambda M$ is continuous, $I^\sim(c) = \chi_c^{\sim -1}(c)$ is closed. $I^\sim(c) = S^1$ clearly is equivalent to $c(t) = c(0)$, all $t \in S$, i.e., $c \in \Lambda^0 M$. The only closed subgroups of S^1, different from S^1, are the cyclic subgroups $\mathbf{Z}_m = \{e^{2\pi i l/m};\ l = 0, \ldots, m-1\}$.

3.A.5 DEFINITION. We call $c \in \Lambda M - \Lambda^0 M$ an element of multiplicity m or m-fold covered if $I^\sim(c) = \mathbf{Z}_m$. If $m = 1$, i.e., if $I^\sim(c) = \mathrm{id}$, then c is also called prime.

The m-fold covering c^m of an element $c \in \Lambda M$ is defined as $c^m(t) = c(mt)$.

REMARK. $I^\sim(c) = \mathbf{Z}_m$ means that $c(t + l/m) = c(t)$, all $t \in S$, all $l \in \mathbf{N}$. For such a c, we can define $c_0(t) = c(t/m)$, $0 \leqslant t \leqslant 1$. c_0 then is called the underlying prime closed curve of c.

3.A.6 DEFINITION. The quotient space $\tilde{\Pi} M$ of ΛM with respect to the S^1-action $\chi^\sim$ is called the space of oriented unparameterized closed curves of M. $\tilde{\Pi} M$ is endowed with the finest topology which makes the projection mapping $\tilde{\pi} : \Lambda M \to \tilde{\Pi} M \equiv \Lambda M/_{\chi^\sim} S^1$ continuous. That is, a subset $B \subset \tilde{\Pi} M$ is open if and only if the counterimage of B under $\tilde{\pi}$ is open.

Note. We also can say that $\tilde{\Pi} M$ is the space of orbits in ΛM under the S^1-action $\chi^\sim$.

Since $E : \Lambda M \to \mathbf{R}$ is constant on the orbits $S^1.c$, we get from E an induced function on $\tilde{\Pi} M$ which we again denote by E. With this we have

3.A.7 THEOREM. *The deformation $\phi_s : \Lambda M \to \Lambda M$ of* 2.13 *induces a deformation $\tilde{\psi}_s : \tilde{\Pi} M \to \tilde{\Pi} M$ such that the following diagram commutes:*

$$\begin{array}{ccc} \Lambda M & \xrightarrow{\ \phi_s\ } & \Lambda M \\ \downarrow{\scriptstyle \tilde{\pi}} & & \downarrow{\scriptstyle \tilde{\pi}} \\ \tilde{\Pi} M & \xrightarrow{\ \tilde{\psi}_s\ } & \tilde{\Pi} M \end{array}$$

PROOF. This follows immediately from grad $E(z.c) = T_c z.\text{grad}\, E(c)$, which is a consequence of 3.A.2. □

3.A.8 DEFINITION. The orientation reversing mapping $\theta : \Lambda M \to \Lambda M$ is defined by $(\theta c)(t) = c(1-t)$. Since $\theta^2 = \text{id}$, this defines a $\mathbf{Z}_2$-action on ΛM.

REMARK. θ may be viewed as stemming from the involution $\theta : \mathbf{R}^2 \to \mathbf{R}^2$; $(x, y) \mapsto (x, -y)$ on $\mathbf{R}^2$, restricted to the canonical embedding $t \in S \mapsto (\cos 2\pi t, \sin 2\pi t) \in \mathbf{R}^2$. When we interpret the S^1-action $\chi^{\sim}$ as stemming from the canonical action of $S\mathbf{O}(2) \cong S^1$ on the circle $S \subset \mathbf{R}^2$, the combined action of S^1 and $\mathbf{Z}_2$ can be viewed as an $\mathbf{O}(2)$-action on ΛM.

3.A.9 LEMMA. *The mapping $\theta : \Lambda M \to \Lambda M$ is an isometry leaving $E : \Lambda M \to \mathbf{R}$ invariant. Moreover, $z.\theta c = \theta(\bar{z}.c)$ for arbitrary $c \in \Lambda M$.*

This shows that θ carries S^1-orbits into S^1-orbits and, thus, θ induces an involution on $\tilde{\Pi} M$ which we again denote by θ. In particular, critical points of E are carried into critical points.

PROOF. The representation of θ in the natural coordinates based at c and θc reads $(\xi(t)) \mapsto (\xi(1-t))$. This shows that θ is an isometry and $E(\theta c) = E(c)$. Finally, if $z = e^{2\pi i r}$,

$$(z.\theta c)(t) = (\theta c)(t + r) = c(1 - t - r) = (\bar{z}.c)(1 - t) = (\theta(\bar{z}.c))(t). \quad \square$$

3.A.10 DEFINITION. The quotient mapping of ΛM under the $\mathbf{Z}_2$-action generated by $\theta : \Lambda M \to \Lambda M$ is denoted by $(\bar{\ }) : \Lambda M \to \bar{\Lambda} M$. $\bar{\Lambda} M$ is called the space of nonoriented parameterized closed curves.

θ induces an involution also on $\tilde{\Pi} M$ which we denote by the same letter. We denote the corresponding quotient mapping by $(\bar{\ }) : \tilde{\Pi} M \to \Pi M$. ΠM is called the space of (nonoriented) unparameterized closed curves. The composition mapping $(\bar{\ }) \circ \tilde{\pi} : \Lambda M \to \Pi M$ will also be denoted by $\pi : \Lambda M \to \Pi M$.

The S^1-action $\chi^{\sim}$ on ΛM determines an equivalence relation on $\bar{\Lambda} M$. We denote the corresponding quotient mapping by $\bar{\tilde{\pi}} : \bar{\Lambda} M \to \Pi M$ and thus get the commutative diagram

$$\begin{array}{ccc} \Lambda M & \xrightarrow{(\bar{\ })} & \bar{\Lambda} M \\ \downarrow{\scriptstyle \tilde{\pi}} & \searrow{\scriptstyle \pi} & \downarrow{\scriptstyle \bar{\tilde{\pi}}} \\ \tilde{\Pi} M & \xrightarrow{(\bar{\ })} & \Pi M \end{array}$$

Note. ΠM can also be viewed as the space of orbits in ΛM under the $\mathbf{O}(2)$-action described above, i.e., the space of orbits under the combined action of S^1 and $\mathbf{Z}_2$.

The following is an immediate consequence of 3.A.9; it constitutes the obvious counterpart to 3.A.7. Note that E is also defined on $\bar{\Lambda} M$ and ΠM.

3.A.11 Theorem. *The deformation* $\phi_s : \Lambda M \to \Lambda M$ *of* 2.13 *induces deformations* $\bar{\phi}_s : \bar{\Lambda} M \to \bar{\Lambda} M$ *and* $\psi_s : \Pi M \to \Pi M$ *such that the following diagrams commute*:

$$\begin{array}{ccc} \Lambda M & \xrightarrow{\phi_s} & \Lambda M \\ \downarrow{\scriptstyle (\bar{\ })} & & \downarrow{\scriptstyle (\bar{\ })} \\ \bar{\Lambda} M & \xrightarrow{\bar{\phi}_s} & \bar{\Lambda} M \end{array}; \qquad \begin{array}{ccc} \Lambda M & \xrightarrow{\phi_s} & \Lambda M \\ \downarrow{\scriptstyle \pi} & & \downarrow{\scriptstyle \pi} \\ \Pi M & \xrightarrow{\psi_s} & \Pi M \end{array}$$

Proof. This follows from $\operatorname{grad}(\theta c) = T_c\theta.\operatorname{grad} E(c)$ and 3.A.7. □

Considering the concepts developed in Chapter 3 for ΛM, we see that most of them are compatible with the S^1-action and the $\mathbf{Z}_2$-action. In particular, we have

3.A.12 Lemma. *The index and the nullity of a critical point* c *of* $E : \Lambda M \to \mathbf{R}$ *are the same for all elements in the* $\mathbf{O}(2)$*-orbit* $S^1.c \cup S^1.\theta c$ *of* c.

Proof. This follows from the canonical isomorphisms between the selfadjoint operators A_c, $A_{z.c}$ and $A_{z.\theta c}$, given by the conjugation with $T_c z$ and $T_c(z\theta)$, respectively. □

We conclude this Appendix by investigating the fixed point set of ΛM under the S^1-action and the $\mathbf{Z}_2$-action described above.

3.A.13 Lemma. *The fixed point set* $\Lambda^0 M$ *of the* S^1*-action, i.e., the set of constant mappings* $c : S \to M$, *is a totally geodesic submanifold of* ΛM *isometric to* M.

Proof. Let $i : M \to \Lambda M$; $p \mapsto c_p$ be the canonical inclusion under which we associate to a point p the constant mapping $c_p : S \to p$. c_p is differentiable. The coordinates of the $c \in \mathfrak{U}(c_p)$ are sections $\xi(t)$ in $c_p^*\tau$ with $\tau^* c_p \xi(t) \in T_p M$. With $\Lambda = \Lambda M$, $\Lambda^0 = \Lambda^0 M$, we define

$$T_{c_p}\Lambda^0 = \left\{\xi_0 \in T_{c_p}\Lambda;\ \xi_0(t) = \text{const}\right\},$$

$$T_{c_p}^{\perp}\Lambda^0 = \left\{\xi \in T_{c_p}\Lambda;\ \langle \xi, \xi_0 \rangle_1 = \langle \xi, \xi_0 \rangle_0 = 0,\ \text{all } \xi_0 \in T_{c_p}\Lambda^0\right\}.$$

Under the canonical identification of $H^1(c_p^*\tau)$ with $T_{c_p}\Lambda M$, the subspace $H^1(\mathcal{O}_{c_p}) \cap T_{c_p}\Lambda^0$ contains the coordinates of the constant curves in $\mathfrak{U}(c_p)$. We therefore have, with $(\exp_{c_p}^{-1}, \mathfrak{U}(c_p))$, a submanifold chart. Thus, $\Lambda^0 = \Lambda^0 M$ is a submanifold.

$\Lambda^0 M$ is totally geodesic since it is the intersection of the fixed point sets of the elements in the 1-parameter family of isometries $z : \Lambda M \to \Lambda M$, $z \in S^1$.

3.A.14 Lemma. *The fixed point set* $\Lambda_\theta M$ *of the involution* $\theta : \Lambda M \to \Lambda M$ *is a totally geodesic submanifold of* ΛM *containing the manifold* $\Lambda^0 M$ *of constant maps. Actually, there exists an isometric embedding*

$$(*) \qquad j : H^1(I, M) \mapsto \Lambda M; \quad c \mapsto jc = \begin{cases} c(2t), & 0 \leqslant t \leqslant \frac{1}{2}, \\ c(2-2t), & \frac{1}{2} \leqslant t \leqslant 1, \end{cases}$$

with $\Lambda_\theta M$ as image. $\Lambda_\theta M$ is transformed into itself under the gradient flow $\phi_s : \Lambda M \to \Lambda M$ on ΛM. There exists a $\rho > 0$ such that a nonconstant critical point of $E : \Lambda M \to \mathbf{R}$ has d_∞-distance $\geqslant \rho$ from the submanifold $\Lambda_\theta M$.

PROOF. Since $\theta : \Lambda = \Lambda M \to \Lambda$ is an isometry, $\Lambda_\theta = \Lambda_\theta M$ is a totally geodesic submanifold.

$c_\theta \in \Lambda_\theta$ means $c_\theta(1-t) = c_\theta(t)$, $t \in S$. In the natural chart based at c_θ, the elements in $\Lambda_\theta \cap \mathfrak{U}(c_\theta)$ are represented by the vector fields $\xi \in H^1(\mathcal{O}_c)$ satisfying

$$(**) \qquad \xi(t) = \xi(1-t).$$

Now, the $\xi \in T_{c\theta}\Lambda$ satisfying $(**)$ clearly form a closed linear subspace of $T_{c\theta}\Lambda$. This again shows that Λ_θ is a submanifold.

The mapping j in $(*)$ is an embedding. To see that j actually is an isometry one need only observe that

$$\langle T_c j.\xi, T_c j.\eta\rangle_1 = \tfrac{1}{2}\langle \xi, \eta\rangle_1 + \tfrac{1}{2}\langle \theta\xi, \theta\eta\rangle_1 = \langle \xi, \eta\rangle_1.$$

$T\theta : T\Lambda \to T\Lambda$ transforms the gradient vector field into itself. Thus, for $c_\theta \in \Lambda_\theta$, $T_{c\theta}\theta.\text{grad}\, E(c_\theta) = \text{grad}\, E(c_\theta)$.

To prove the last statement we consider the injectivity radius $\iota(M)$ of M. On every closed geodesic c' we have a parameter t_0, $0 < t_0 \leqslant \frac{1}{4}$ such that $d(c'(t_0), c'(1-t_0)) = \iota(M)$. Then, if $c_\theta \in \Lambda_\theta$,

$$\begin{aligned} \iota(M) &= d(c'(t_0), c'(1-t_0)) \\ &\leqslant d(c'(t_0), c_\theta(t_0)) + d(c_\theta(1-t_0), c'(1-t_0)) \leqslant 2d_\infty(c', c_\theta); \end{aligned}$$

cf. 1.1 for the definition of d_∞. Thus, $\rho = \iota(M)/2$ is a positive number such that $d_\infty(c', \Lambda_\theta) \geqslant \rho$ whenever c' is a nonconstant critical point of E. This also implies a positive bound for $d_\Lambda(c', \Lambda_\theta)$, since d_Λ is finer than d_∞ (cf. 3.6). Actually, it is easy to establish the relation $d_\infty(c_1, c_2) \leqslant \sqrt{2}\, d_\Lambda(c_1, c_2)$ for c_1, c_2 in Λ (cf. Klingenberg [3]). □

REMARK. As we saw in 2.1 and 2.2, ΛM is a canonically embedded submanifold of $H^1(I, M)$. On the other hand, $H^1(I, M)$ possesses an isometric embedding into ΛM. We have here an example of a Riemannian Hilbert manifold, i.e., $H^1(I, M)$, which can be embedded isometrically into itself. As a submanifold, it has infinite codimension. Clearly, a Riemannian manifold of finite dimension can never be embedded isometrically into itself with codimension > 0.

Other examples can easily be given when a Hilbert manifold can be embedded into itself with codimension > 0. Take, e.g., the mapping $m : c \in \Lambda M \mapsto c^m \in \Lambda M$, m an integer > 1. In this case, the embedding is not isometric, however.

We conclude with a remark on the problem of the existence of closed geodesics which are geometrically distinct. By this we mean the following: First, it means that the geodesics should not just differ by parameterization, i.e., they should represent different elements in ΠM. In addition, however, we want geometrically distinct closed geodesics not to have the same underlying prime closed geodesics.

Note that every closed geodesic c gives rise to a whole tower $\{c^m, m = 1, 2, \dots\}$ of closed geodesics, i.e., the series of its multiple coverings. Two different elements in the same tower should not be viewed as geometrically distinct.

In other words, finding geometrically distinct closed geodesics amounts to finding prime closed geodesics which represent different elements in ΠM. As we mentioned earlier, we must refer the reader to Klingenberg **[3]** for an exposition of the methods and the results in the problem of finding many geometrically distinct closed geodesics. Cf. also Chapter 5.

Chapter 4. Closed geodesics on spheres

In 2.20 we proved the existence of a closed geodesic on a compact Riemannian manifold. We now develop additional methods which will be employed in the construction of several closed geodesics.

The principal tool is the space of circles on a sphere S^n. A nonnull homotopic mapping of S^n into a manifold M can sometimes be used to prove the existence of a good number of closed geodesics on M. The main results concern the case where M is homeomorphic to S^n. We begin with the space of circles on S^n with $n \geqslant 2$. We think of S^n as embedded the standard way into the Euclidean space $\mathbf{R}^{n+1}$, i.e., $S^n = \{\Sigma_0^n x_i^2 = 1\}$.

4.1 DEFINITION. (i) The space $\mathrm{A}S^n$ of parameterized circles on S^n consists of the mappings $c: S \to S^n$ where c is either a constant map—in which case we also call c a point circle—or an embedding of the circle $S = [0,1]/\{0,1\}$ with parameter proportional to arc length. The image shall lie in the intersection of S^n and a 2-plane in $\mathbf{R}^{n+1}$ having distance < 1 from the origin $0 \in \mathbf{R}^{n+1}$.

We endow $\mathrm{A}S^n$ with the topology induced from the inclusion $\mathrm{A}S^n \hookrightarrow \Lambda S^n$.

(ii) A parameterized great circle is a circle where the image lies in a 2-plane through $0 \in \mathbf{R}^{n+1}$. That is, it is a simple closed geodesic on S^n with parameter interval $[0, 1]$. By $\mathrm{B}S^n$ we denote the space of great circles.

4.2 PROPOSITION. *The space* $\mathrm{B}S^n$ *of parameterized great circles is canonically isomorphic to the unit tangent bundle* $T_1 S^n$. *The standard action of* $\mathbf{O}(n+1)$ *on* S^n *operates transitively on* $\mathrm{B}S^n$ *with isotropy group isomorphic to* $\mathbf{O}(n-1)$.

PROOF. A great circle $c(t)$, $0 \leqslant t \leqslant 1$, is determined by $X_0 = \dot{c}(0)/|\dot{c}(0)| \in T_1 M$, and vice verse. Thus, $\mathrm{B}S^n \cong T_1 S^n$.

It is known that $S^n = \mathbf{O}(n+1)/\mathbf{O}(n)$. Consider, on S^n, the point $p_0 = (1,0,\ldots,0)$ and, in $T_{p_0}S^n$, the vector X_0 with coordinates $(0,1,\ldots,0)$. The subgroup of $\mathbf{O}(n+1)$, which leaves X_0 invariant, is the group $\mathbf{O}(n-1) \subset \mathbf{O}(n+1)$, which operates as the identity on the (x_0, x_1)-plane of $\mathbf{R}^{n+1}$ and in the standard way on the complementary orthogonal $(n-1)$-dimensional subspace.

4.3 PROPOSITION. *The space* $\mathrm{A}S^n - \mathrm{A}^0 S^n$ *of nonconstant circles on* S^n *is the total space of a* D^{n-1}*-bundle* $\alpha: \mathrm{A}S^n - \mathrm{A}^0 S^n \to \mathrm{B}S^n$ *over* $\mathrm{B}S^n$. *Here, the image under* α *of a circle* c *is the parameterized great circle* c^* *which is obtained by first parallel translating the plane of* c *into the origin* $0 \in \mathbf{R}^{n+1}$ *such that the midpoint of* c *goes into* 0 *and then blowing up the circle into a great circle* c^*.

Note. When we speak of a D^{n-1}-bundle we mean the restriction of a vector bundle with fibre dimension $(n-1)$ to the open unit discs in each fibre.

PROOF. Obviously, the mapping α is well determined and surjective. The counterimage $\alpha^{-1}(c^*)$ of a great circle c^*, contained in the 2-plane $\mathbf{R}^{*2} \subset \mathbf{R}^{n+1}$, can be described by the set of midpoints of the circles c with $\alpha(c) = c^*$. The midpoints have distance < 1 from $0 \in \mathbf{R}^{n+1}$ and belong to the $(n-1)$-dimensional space $(\mathbf{R}^{*2})^{\perp}$ in $\mathbf{R}^{n+1}$ which is orthogonal to $\mathbf{R}^{*2}$. That is, $\alpha^{-1}(c^*)$ can be identified with the unit disc in $(\mathbf{R}^{*2})^{\perp}$. α is thus seen to be the restriction of an $(n-1)$-dimensional vector bundle. □

Recall that we consider $\mathrm{A}S^n$, A^0S^n and $\mathrm{B}S^n$ as subspaces of ΛS^n with the induced topology. All these spaces are transformed into themselves by the involution $\theta : \Lambda M \to \Lambda M$ as well as by the S^1-action $\chi^{\sim}$ defined in the Appendix to Chapter 3. The involution θ, given by $(c(t)) \mapsto (c(1-t))$, and the involution consisting of the action of $e^{i\pi} \in S^1$, i.e., $(c(t+1/2))$, commute with each other. We put $e^{i\pi}\theta = \theta e^{i\pi} = \theta'$. This leads us to consider the following quotient spaces of $\mathrm{A}S^n$, A^0S^n, $\mathrm{B}S^n$:

4.4 DEFINITION. (i) The quotient spaces of $\mathrm{A}S^n$, A^0S^n, $\mathrm{B}S^n$ under the $\mathbf{Z}_2$-action generated by θ are denoted by $\overline{\mathrm{A}}S^n$, $\overline{\mathrm{A}}^0S^n$, $\overline{\mathrm{B}}S^n$, respectively. We view them as subspaces of $\overline{\Lambda}S^n = \Lambda S^n/_{\theta}\mathbf{Z}_2$.

(ii) The quotient spaces of $\mathrm{A}S^n$, A^0S^n, $\mathrm{B}S^n$ under the $\mathbf{Z}_2$-action generated by θ' are denoted by A^*S^n, $\mathrm{A}^{*0}S^n$, B^*S^n, respectively. They are subspaces of $\Lambda^*S^n = \Lambda S^n/_{\theta'}\mathbf{Z}_2$.

(iii) The quotient spaces of $\mathrm{A}S^n$, A^0S^n, $\mathrm{B}S^n$ under the $(\mathbf{Z}_2 \times \mathbf{Z}_2)$-action generated by θ, θ' are denoted by $\overline{\mathrm{A}}^*S^n$, $\overline{\mathrm{A}}^{*0}S^n$, $\overline{\mathrm{B}}^*S^n$, respectively. We view them as subspaces of $\overline{\Lambda}^*S^n = \Lambda S^n/_{(\theta,\theta')}(\mathbf{Z}_2 \times \mathbf{Z}_2)$.

REMARK. Traditionally (cf. Klingenberg [3]) one also considers the quotient spaces ΓS^n, $\Gamma^0 S^n$, δS^n of $\mathrm{A}S^n$, A^0S^n, $\mathrm{B}S^n$ with respect to the full $\mathbf{O}(2)$-action χ. ΔS^n is then called the space of unparameterized great circles. Such a great circle can be identified with the 2-plane through $0 \in \mathbf{R}^{n+1}$, which is the carrier of the great circles. Thus, ΔS^n can be identified with the Grassmann manifold $G(2, n-1)$ of 2-planes through $0 \in \mathbf{R}^{n+1}$.

4.5 PROPOSITION. (i) *The space* $\mathrm{B}S^n$ *of parameterized great circles on* S^n *corresponds to the Stiefel manifold* $V(2, n-1)$ *of pairs* $\{e, e'\}$ *of orthonormal vectors in* $\mathbf{R}^{n+1}$. *Take for* e *the initial point of the great circle and for* e' *its initial tangent vector normed to the length* 1. *Thus,* $\mathrm{B}S^n$ *can be identified with the homogeneous space*

$$\mathbf{O}(n+1)/\mathbf{O}(n-1) \cong \mathbf{O}(n+1)/S\mathbf{O}(1) \times S\mathbf{O}(1) \times \mathbf{O}(n-1).$$

Here we have indicated on the right-hand side the way in which $\mathbf{O}(n-1)$ *is embedded in* $\mathbf{O}(n+1)$.

(ii) *The involutions* θ', θ *generate the subgroups* $\mathbf{O}(1) \times S\mathbf{O}(1) \times \mathrm{id}_{n-1}$ *and* $S\mathbf{O}(1) \times \mathbf{O}(1) \times \mathrm{id}_{n-1}$ *of* $\mathbf{O}(n+1)$, *while the* $\mathbf{O}(2)$-*action on* $\mathrm{B}S^n$ *is given by*

$\mathbf{O}(2) \times \mathrm{id}_{n-1}$. *Hence,*

$$\overline{\mathrm{B}}S^n \cong \mathbf{O}(n+1)/S\mathbf{O}(1) \times \mathbf{O}(1) \times \mathbf{O}(n-1),$$
$$\mathrm{B}^*S^n \cong \mathbf{O}(n+1)/\mathbf{O}(1) \times S\mathbf{O}(1) \times \mathbf{O}(n-1),$$
$$\overline{\mathrm{B}}^*S^n \simeq \mathbf{O}(n+1)/\mathbf{O}(1) \times \mathbf{O}(1) \times \mathbf{O}(n-1),$$
$$\Delta S^n \simeq \mathbf{O}(n+1)/\mathbf{O}(2) \times \mathbf{O}(n-1).$$

(iii) *Since the projection mapping α commutes with the S^1- and $\mathbf{Z}_2$-actions, we get the D^{n-1}-bundles*

$$\bar{\alpha} : \overline{\Lambda} S^n - \overline{\Lambda}^0 S_n \to \overline{\mathrm{B}}S^n,$$
$$\alpha^* : \Lambda^* S^n - \Lambda^{*0} S^n \to \mathrm{B}^*S^n,$$
$$\bar{\alpha}^* : \overline{\Lambda}^* S^n - \overline{\Lambda}^{*0} S^n \to \overline{\mathrm{B}}^*S^n,$$
$$\gamma : \Gamma S^n - \Gamma^0 S^n \to \Delta S^n.$$

PROOF. Clear from the definitions. □

Note. $\mathrm{B}S^n$ can be viewed in two ways as the total space of an S^{n-1}-bundle over S^n,

$$S^{n-1} \to \mathrm{B}S^n \xrightarrow{\tau} S^n; \quad S^{n-1} \to \mathrm{B}S^n \xrightarrow{\sigma} S^n,$$

by associating with $\{e, e'\} \in \mathrm{B}S^n \cong V(2, n-1)$ the element $e \in S^n$ or $e' \in S^n$, respectively. θ and θ' operate as involutions on the fibre of τ and σ, respectively. Taking the quotient space of these actions, we obtain the projective bundles

$$P^{n-1} \to \overline{\mathrm{B}}S^n \xrightarrow{\bar{\tau}} S^n; \quad P^{n-1} \to \mathrm{B}^*S^n \xrightarrow{\sigma^*} S^n,$$

which are clearly isomorphic to each other.

One can show that the $\mathbf{Z}_2$-cohomology ring of $\overline{\mathrm{B}}S^n \cong \mathrm{B}^*S^n$ is given by

$$H^*(\overline{\mathrm{B}}S^n) \cong H^*(P^{n-1}) \otimes H^*(S^n).$$

Indeed, the $\mathbf{Z}_2$-action θ on $\mathrm{B}S^n$ is a fixed point free isometry, i.e., $\mathrm{B}S^n$ is a 2-fold covering of $\overline{\mathrm{B}}S^n$. The corresponding 1-dimensional cohomology class is the generator of the ring $H^*(P^{n-1}) = \mathbf{Z}_2[X]/(X^n)$.

We use the n cycles in the fibre P^{n-1} of $\overline{\mathrm{B}}S^n$ to define n ϕ-families in $\overline{\Lambda}^*S^n$. We begin with

4.6 DEFINITION. Denote by B_k, $0 \leqslant k \leqslant n-1$, the subset of $\mathrm{B}S^n$ formed by those parameterized great circles on S^n which start at $p_0 = (1, 0, \ldots, 0) \in S^n$ and lie on

$$S^{k+1} = \left\{ \sum_0^{k+1} x_i^2 = 1 \right\}.$$

Clearly B_k is θ-invariant.

The restriction of the D^{n-1}-bundle α to $\mathrm{B}_k \subset \mathrm{B}S^n$ is denoted by $\alpha_k : \Lambda_k - \Lambda_k^0 \to \mathrm{B}_k$. Denote by Λ_k, $\overline{\Lambda}_k$ the closure in ΛS^n, $\overline{\Lambda}^*S^n$ of $\Lambda_k - \Lambda_k^0$, $\overline{\Lambda}_k - \overline{\Lambda}_k^0$. (Λ_k, Λ_k^0), or simply Λ_k, represents a $(k+n-1)$-chain in $(\Lambda S^n, \Lambda^0 S^n)$. $(\overline{\Lambda}_k, \overline{\Lambda}_k^0)$ is a $(k+n-1)$-cycle in $(\overline{\Lambda} S^n, \overline{\Lambda}^0 S_n)$.

REMARK. B_k can be identified with the sphere S^k and $\overline{\mathrm{B}}_k$ can be identified with P^k, a generator of the k-dimensional $\mathbf{Z}_2$-homology of the fibre P^{n-1} of $\bar{\tau}$ over p_0. Thus, the $(k+n-1)$-cycle $(\overline{\mathrm{A}}_k, \overline{\mathrm{A}}_k^0)$ can be viewed as a 'thickening' of the base cycle $\overline{\mathrm{B}}_k$ by the D^{n-1}-discs over $\overline{\mathrm{B}}_k$ in the bundle $\bar{\tau}_k$ modulo its boundary.

We use this to define ϕ-families in the sense of 2.17.

4.7 DEFINITION. Let M be a Riemannian manifold for which there exists a homeomorphism $\phi: S^n \to M$ with the property that it induces a continuous mapping of the space of circles ΛS^n into ΛM.

(i) Denote by A_k any chain in $(\Lambda S^n, \Lambda^0 S^n)$ which is θ-equivariantly homologous to the chain A_k defined in 4.6.

(ii) Let $u_k: \mathrm{A}_k \to (\Lambda M, \Lambda^0 M)$ be the θ-equivariant mapping induced by ϕ. Thus, by going to the quotient modulo the $\mathbf{Z}_2$-action θ, we get a $\mathbf{Z}_2$-cycle $\bar{u}_k: \overline{\mathrm{A}}_k \to (\overline{\Lambda} M, \overline{\Lambda}^0 M)$. We define the ϕ-family $\mathcal{Q}_k$ so as to consist of the sets $\phi_s u_k(\mathrm{A}_k)$. Since ϕ_s commutes with the $\mathbf{Z}_2$-action θ, this is indeed a ϕ-family.

Note. The hypothesis on M holds for every manifold for which the underlying topological manifold is the sphere S^n. One way to see this is that on the n-sphere every differentiable structure is related to the standard differentiable structure by a homeomorphism which is differentiable, with the possible exception of a single point (cf. Milnor [**1**]). Another way would be to observe that any continuous mapping $\phi: S^n \to M$ is homotopically equivalent to a differentiable $\phi: S^n \to M$ (cf. Hirsch [**1**]).

4.8 THEOREM. *Let M be a manifold satisfying the hypothesis in* 4.7. *Denote by κ_k the critical values of the ϕ-families $\mathcal{Q}_k$, $0 \leq k \leq n-1$. Then $0 < \kappa_0 \leq \kappa_1 \leq \cdots \leq \kappa_{n-1}$. If $\kappa_l = \kappa_{l+1} =$ (briefly) κ for some $l < n-1$, then the compact set of critical points with E-value κ consists of an infinite number of S^1-orbits.*

REMARKS. 1. This result essentially goes back to Lyusternik [**1**]. Lyusternik, and later Alber [**1**], considered the space $(\Gamma S^n, \Gamma^0 S^n)$ of unparameterized circles instead of the space $(\overline{\Lambda} S^n, \overline{\Lambda}^0 S^n)$.

The advantage of taking the cycles $\bar{u}_j$ of $\overline{\Lambda} S^n \bmod \overline{\Lambda}^0 S^n$ instead of the cycles $v_j = \tilde{\pi}(\bar{u}_j)$ of $\Gamma S^n \bmod \Gamma^0 S^n$ lies in the fact that $\overline{\Lambda} S^n$ is closer to the manifold ΛS^n than ΠS^n is. Actually, if we remove from $\overline{\Lambda} S^n$ the set $\overline{\Lambda}_\theta S^n$, we again get a manifold having $\Lambda S^n - \Lambda_\theta S^n$ as its double covering (cf. 3.A.14). Alber's claim that the quotient cycles $v_j = \tilde{\pi}(\bar{u}_j)$ in $(\Pi S^n, \Pi^0 S^n)$ are nonhomologous to zero is not correct. However, one does not need this as our proof will show.

2. 4.8 can be generalized to the case when we have a mapping $\phi: S^l \to M$, $2 \leq l \leq \dim M$, which defines a nontrivial homology class (cf. Klingenberg [**3, 4**]).

PROOF. From the standard metric on S^n it follows that $H_i(\overline{\Lambda} S^n, \overline{\Lambda}^0 S^n) = 0$ for $i < n-1$, while according to 2.20, $\pi_{n-1}(\overline{\Lambda} S^n, \overline{\Lambda}^0 S^n) \neq 0$. Note that the restriction to θ-invariant homotopies does not invalidate the arguments given there. Hence, by the Hurewicz isomorphism (cf. Spanier [**1**]), $H_{n-1}(\overline{\Lambda} S^n, \overline{\Lambda}^0 S^n) = 0$, which shows $\kappa_0 > 0$. $\kappa_k \leq \kappa_{k+1}$ follows from the fact that a chain u_{k+1} contains, by restriction, a chain u_k.

The only case that remains to be discussed is $\kappa_l = \kappa_{l+1} =$ (briefly) κ for some $l < n - 1$. Denote by Cr κ the set of critical points with E-value κ. We will derive a contradiction from the assumption that Cr κ consists of only finitely many pairs of critical orbits $(S^1.c_\iota, S^1.\theta c_\iota)$, $\iota \in I$, I some finite set. To do this we choose for each $\iota \in I$ an S^1-invariant neighborhood $\mathcal{U}_\iota$ of $S^1.c_\iota$ outside the fixed point set $\Lambda_\theta M$ of $\theta : \Lambda M \to \Lambda M$. We can assume $\mathcal{U}_\iota \cap \theta\mathcal{U}_\iota = \varnothing$, all $\iota \in I$, and $\mathcal{U}_\iota \cap \mathcal{U}_\kappa = \mathcal{U}_\iota \cap \theta\mathcal{U}_\kappa = \varnothing$, all $(\iota, \kappa) \in I \times I$, $\iota \neq \kappa$. By $\mathcal{U}$ we denote the union of these $\mathcal{U}_\iota, \theta\mathcal{U}_\iota$, $\iota \in I$. Then $\theta\mathcal{U} = \mathcal{U}$. If we denote by $\overline{\mathcal{U}}$ the image of $\mathcal{U}$ in $\overline{\Lambda} M$, then $\overline{\mathcal{U}} \subset \overline{\Lambda} M - \overline{\Lambda}_\theta M$. On $\overline{\Lambda} M - \overline{\Lambda}_\theta M$ we have the 1-dimensional $\mathbf{Z}_2$-homology class which is represented by the image in $\overline{\Lambda} M - \overline{\Lambda}_\theta M$ of any curve in $\Lambda M - \Lambda_\theta M$ from some c to θc. $\overline{\mathcal{U}}$ does not carry a cycle in this class.

There exists a $u_{l+1} : A_{l+1} \to \Lambda M$ with image in $\mathcal{U} \cup \Lambda^{\kappa -} M$ (cf. 2.18). Put $u_{l+1}^{-1}(\mathcal{U}) = \mathcal{O}$ and $u_{l+1}^{-1}(\Lambda^{\kappa -} M) = \mathcal{N}$. $\mathcal{O} \subset \mathrm{A}_{l+1} - \mathrm{A}^0_{l+1}$. $\mathcal{O}$ and $\mathcal{N}$ are θ-invariant. $\mathcal{O}$ does not carry a curve from some c_0 to θc_0. That is, $\overline{\mathcal{O}}$ does carry a cycle representing the 1-dimensional homology class = homotopy class of $\overline{\mathrm{A}}_{l+1} - \overline{\mathrm{A}}^0_{l+1} \cong P^{l+1}$. To put it differently, every nonnull homotopic curve in $\overline{\mathrm{A}}_{l+1} - \overline{\mathrm{A}}^0_{l+1} \cong P^{l+1}$ has points outside $\overline{\mathcal{O}}$. Therefore there exists a cycle $\overline{\mathrm{A}}_l \subset \overline{\mathcal{N}}$. But then $u_l(\mathrm{A}_l) \subset \Lambda^{\kappa -} M$, i.e., $\kappa_l < \kappa_{l+1}$. □

REMARKS. 1. Another possibility is to use the cap product between the $(l + n)$-dimensional cycle $\overline{\mathrm{A}}_{l+1}$ and a 1-dimensional nontrivial cocycle ω on $\overline{\mathrm{A}}_{l+1} - \overline{\mathrm{A}}^0_{l+1} \cong \overline{\mathrm{B}}_{l+1} \cong P^{l+1}$. This yields a cycle $\overline{\mathrm{A}}_l$ inside $\overline{\mathcal{N}}$. We come back to this in the proof of 4.12.

2. Unless $\kappa_l = \kappa_{l+1}$ for some $l < n - 1$, 4.8 does not imply the existence of n different prime (unparameterized) closed geodesics on a manifold M of the homotopy type of S^n. Indeed, we have not excluded the possibility that among the critical orbits at the n different E-levels $\kappa_0 < \kappa_1 < \cdots < \kappa_{n-1}$ some, or even all, are just multiple coverings of the same underlying prime critical orbit. That this actually is not the case, and that the underlying prime closed geodesics of the n closed geodesics with E-value $\kappa_0 < \cdots < \kappa_{n-1}$ are all different has been shown in Klingenberg [**3**] with the help of the so-called Morse complex. Here we prove an even stronger result under the additional assumption that the sectional curvature of M is more than $\frac{1}{4}$-pinched. The original proof was given in Klingenberg [**1**].

4.9 THEOREM. *Let M be a simply connected compact manifold for which the sectional curvature K satisfies the relation $K_1/4 < K \leqslant K_1$ for some $K_1 > 0$. M is then homeomorphic to S^n, $n = \dim M$. There exist on M n different simple closed unparameterized geodesics with length in the interval* $[2\pi/\sqrt{K_1}, 4\pi/\sqrt{K_1}[$.

PROOF. By normalizing the metric, we can restrict ourselves to the case $K_1 = 1$, i.e., $K_0 \leqslant K \leqslant 1$, with some $K_0 > \frac{1}{4}$. From the proof of the sphere theorem we have the existence of a homeomorphism $\phi : S^n \to M$, $n = \dim M$, with the following properties:

Let (p, q) be points on M with maximal distance $d(p, q) = d(M)$. Then $\pi \leq d(M) \leq \pi\sqrt{K_0} < 2\pi$. The set $E = E(p, q)$ of points $r \in M$ with $d(p, r) = d(q, r)$ is an embedding of the $(n-1)$-sphere S^{n-1}. We also call E the equator of M w.r.t. (p, q). $d(p, E) \geq \pi/2$, $d(q, E) > \pi/2$. $M - E$ consists of two disjoint open sets $B(p)$ and $B(q)$ with $\partial\bar{B}(p) = \partial\bar{B}(q) = E$. The open ball $B_\pi(p)$ of radius π around p is entirely in the domain of polar coordinates based at p. $\bar{B}(p) = B(p) \cup E$ belongs to $B_\pi(p)$ and every geodesic emanating from p meets $\partial\bar{B}(p) = E$ transversally. The corresponding facts hold for $B(q)$.

Denote by M_1 the sphere S^n of constant curvature $K_1 = 1$. On M_1 we choose $p_1 = (1, 0, \dots, 0)$ and $q_1 = (-1, 0, \dots, 0)$. Define $E_1 = E(p_1, q_1)$, $B(p_1)$, $B(q_1)$ as before. Hence, $B(p_1) = B_{\pi/2}(p_1)$, $B(q_1) = B_{\pi/2}(q_1)$. The mapping $\phi \colon M_1 = S^n \to M$, constructed in the proof of the sphere theorem, now satisfies the hypotheses necessary for the proof of 4.9.

In particular, with a half great circle $c_1(t)$, $0 \leq t \leq 1$, from p_1 to q_1, there is associated a curve $c(t) = \phi \circ c_1(t)$, $0 \leq t \leq 1$, from p to q where $c \mid [0, \frac{1}{2}]$ is the minimizing geodesic from p to a point r on E, and $c \mid [\frac{1}{2}, 1]$ is the minimizing geodesic from r to q. $L(c) < 2\pi$.

With the help of ϕ we define a cycle $u_k \colon (\mathrm{A}_k, \mathrm{A}_k^0) \to (\Lambda M, \Lambda^0 M)$ of the type considered in 4.7 by associating with a circle $c_0(t)$, $0 \leq t \leq 1$, in A_k the closed curve $\phi \circ c_0(t)$, $0 \leq t \leq 1$. The image of a great circle through (p_1, q_1) has length $< 4\pi$; its E-value is $< 8\pi^2$. We will show u_{n-1} can be deformed θ-equivariantly such that its image is entirely below the E-level $< 8\pi^2$.

Once this is done we can complete the proof of the theorem as follows: First, $\kappa_{n-1} < 8\pi^2$, i.e., all closed geodesics constructed with the help of the ϕ-families $\mathcal{Q}_k$, $0 \leq k \leq n-1$, defined in 4.7 have length $< 4\pi$. On the other hand, a nonconstant geodesic loop has length $\geq 2\pi$. Thus, our closed geodesics are all simple. 4.8 then shows that there must be at least n simple closed geodesics.

There remains to define the θ-equivariant deformation of u_{n-1} into some $\tilde{u}_{n-1}(\mathrm{A}_{n-1}) \subset \Lambda^{8\pi^2-}M$. To do this observe that each circle in A_{n-1} is completely determined by its intersection with the equator $E_1 = E(p_1, q_1)$ on S^n. This intersection consists of exactly two points, except when the circle is a constant curve, i.e., except when the circle belongs to A_{n-1}^0, in which case it consists of a single point on E_1. Now replace each circle of $\mathrm{A}_{n-1} - \mathrm{A}_{n-1}^0$ by the closed curve formed by the half great circle from p_1 to q_1 which passes through the first of the two intersection points of the circle with the equator E_1 and then go back to p_1 through the second intersection point. Replace a degenerate circle in A_{n-1}^0 by the half great circle from p_1 to q_1, passing through that degenerate circle, and back the same way. Finally, add the family of closed curves obtained by retracting the latter curves back into their intersection with the equator. The newly defined set $\tilde{\mathrm{A}}_{n-1} \subset \Lambda S^n$ is clearly θ-equivariantly homotopic to A_{n-1}. For every $\tilde{c} \in \tilde{\mathrm{A}}_{n-1}$, its image $\phi \circ \tilde{c} \in \Lambda M$ consists of closed curves which are formed by not more than four geodesic segments, each of which has length less than the distance from p (or from q) to a point r on the equator E. Since $d(p, r) = d(q, r) < \pi$, the curves $\phi \circ \tilde{c}$ all have E-value $< 8\pi^2$. Now use the θ-equivariant homotopy from A_{n-1} into $\tilde{\mathrm{A}}_{n-1}$ to define a θ-equivariant homotopy from u_{n-1} into an $\tilde{u}_{n-1}$ with image in $\Lambda^{8\pi^2-}M$. $\square$

We wish to define a larger set of ϕ-families than the one defined in 4.7. For that purpose we take into consideration not only the $\mathbf{Z}_2$-action generated by θ but also the one generated by $\theta' = e^{i\pi}\theta$. Thus, we consider the D^{n-1} bundle $\bar{\alpha}^*$ from 4.5. For such a bundle we have the Thom isomorphism

$$H_*(\overline{\mathrm{B}}^*S^n) \cong H_{*+(n-1)}(\overline{\mathrm{A}}^*S^n, \overline{\mathrm{A}}^{*0}S^n),$$

which is defined as follows (cf. Spanier [**1**]): If $\bar{y}_k^*$ is a k-cycle in $\overline{\mathrm{B}}^*S^n$, take the closure $\bar{z}_k^*$ of the fibres $\bar{\alpha}^{*-1}(\bar{y}_k^*)$ over $\bar{y}_k^*$; this is a relative $(k+n-1)$-cycle of $(\overline{\mathrm{A}}^*S^n, \overline{\mathrm{A}}^{*0}S^n)$.

Besides the D^{n-1}-bundle $\bar{\alpha}^*$, we also consider the D^{n-1}-bundle γ over ΔS^n (cf. 4.5). Here again we have the Thom isomorphism

$$H_*(\Delta S^n) \cong H_{*+(n-1)}(\Gamma S^n, \Gamma^0 S^n).$$

We also need the ring structures of $H^*(\overline{\mathrm{B}}^*S^n)$ and $H^*(\Delta S^n)$. For the latter case it was determined by Chern [**1**]. For our purposes it is convenient to also use the description given by Borel [**1**], which is valid for a general class of homogeneous spaces. According to Borel we can write

$$H^*(\Delta S^n) = S(v_1, v_2) \otimes S(v_3, \ldots, v_{n+1})/S^+(v_1, v_2, v_3, \ldots, v_{n+1}).$$

Here, $S(v_1, \ldots, v_k)$ denotes the ring of symmetric polynomials in $(v_1, \ldots, v_k)$ with $\mathbf{Z}_2$-coefficients, and $S^+(v_1, \ldots, v_k)$ is the ideal in $S(v_1, \ldots, v_k)$ formed by the polynomials of degree > 0. An additive basis for $H^*(\Delta S^n)$ is given by the elements

$$(a, b) = \sum_0^{b-a} v_1^{a+i} \cdot v_2^{b-i}, \qquad 0 \leq a \leq b \leq n-1.$$

The multiplicative structure of $H^*(\Delta S^n)$ is determined by the rules

$$(a_1, a_2) = (0, a_1) \cup (0, a_2) + (0, a_1 - 1) \cup (0, a_2 + 1),$$
$$(0, a) \cup (a_1, a_2) = \sum_i (a_1 + i, a_2 + a - i); \qquad 0 \leq i \leq \min(a, a_2 - a_1).$$

Here we put $(a, b) = 0$, whenever $0 \leq a \leq b \leq n-1$ is not satisfied (cf. Chern [**1**]).

We also need $H^*(\overline{\mathrm{B}}^*S^n)$ which, according to Borel [**1**], can be written as

$$H^*(\overline{\mathrm{B}}^*S^n) = S(u_1) \otimes S(u_2) \otimes S(u_3, \ldots, u_{n+1})/S^+(u_1, u_2, u_3, \ldots, u_{n+1}).$$

If $\pi : \overline{\mathrm{B}}^*S^n \to \Delta S^n$ is the canonical projection given by the quotient mapping of the S^1-action, then the induced homomorphism $\pi^* : H^*(\Delta S^n) \to H^*(\overline{\mathrm{B}}^*S^n)$ is an embedding, where the image is formed by the subring of polynomials symmetric in u_1, u_2—note that all elements of the representation of $H^*(\Delta S^n)$ can be expressed by v_1, v_2 alone. In particular,

$$\pi^*(v_1 + v_2) \equiv \pi^*(0,1) = u_1 + u_2;$$
$$\pi^*(v_1^2 + v_1 \cdot v_2 + v_2^2) \equiv \pi^*(0,2) = u_1^2 + u_1 \cdot u_2 + u_2^2;$$
$$\pi^*(v_1 \cdot v_2) = \pi^*(1,1) = u_1 \cdot u_2.$$

Alber [**1**] has determined the maximal number of cohomology classes of ΔS^n, all with the exception of one of positive dimension, which together have a product $\neq 0$. If we write $n = 2^k + s$ with $0 \leqslant s < 2^k$, such a product is given by

$$1 \cdot (v_1 + v_2)^{2n-2s-2} \cdot (v_1 \cdot v_2)^s = v_1^{n-1} \cdot v_2^{n-1}$$

or, using our second notation,

$$1 \cdot (0,1)^{2n-2s-2} \cdot (1,1)^s = (n-1, n-1).$$

To see this observe that $(v_1 + v_2)^{2^l} = v_1^{2^l} + v_2^{2^l}$. Hence

$$(v_1 + v_2)^{2^k - 1} = \sum_0^{2^k - 1} v_1^i \cdot v_2^{2^k - i - 1},$$

and

$$(v_1 + v_2)^{2^{k+1} - 2} = \left((v_1 + v_2)^{2^k - 1}\right)^2 = \sum_i v_1^{2i} \cdot v_2^{2^{k+1} - 2i - 2} \neq 0$$

since there occurs the summand $v_1^{2^k - 2} \cdot v_2^{2^k} + v_1^{2^k} \cdot v_2^{2^k - 2}$. But $(v_1 + v_2)^{2^{k+1} - 1} = 0$ since $2^{k+1} - 1 > 2n - 2$. Multiplying the last equation by $(v_1 \cdot v_2)^s$ yields $v_1^{n-1} \cdot v_2^{n-1}$.

We consider the cap product

$$H_*(\bar{\Lambda}^*S^n, \bar{\Lambda}^{*0}S^n) \otimes H^*(\bar{B}^*S^n) \xrightarrow{\cap} H_*(\bar{\Lambda}^*S^n, \bar{\Lambda}^{*0}S^n).$$

On the level of simplices and cochains, this is defined as follows (cf. Spanier [**1**]): for a $(j+k+n-1)$-simplex $\bar{s}^*_{j+k}$ of $\bar{\Lambda}^*S^n$ and a k-cochain ω^k of $\bar{B}^*S^n$, $\bar{s}^*_j = \bar{s}^*_{j+k} \cap \omega^k$ is the front $(j+n-1)$-simplex of $\bar{s}^*_{j+k}$ with coefficient given by the value of ω^k on the back k-simplex of $\bar{s}^*_{j+k}$. Thus, a cohomology class $\omega^k \in H^k(\bar{B}^*S^n)$ defines a mapping

$$H_{j+k+n-1}(\bar{\Lambda}^*S^n, \bar{\Lambda}^{*0}S^n) \xrightarrow{\cap \omega^k} H_{j+n-1}(\bar{\Lambda}^*S^n, \bar{\Lambda}^{*0}S^n).$$

Similarly, we have the cap product

$$H_*(\Gamma S^n, \Gamma^0 S^n) \oplus H^*(\Delta S^n) \xrightarrow{\cap} H_*(\Gamma S^n, \Gamma^0 S^n).$$

Above, we wrote down a basis $\{(a,b); 0 \leqslant a \leqslant b \leqslant n-1\}$ for $H^*(\Delta S^n)$. Denote the dual basis for $H_*(\Delta S^n)$ by $\{[a,b]; 0 \leqslant a \leqslant b \leqslant n-1\}$. We also use the symbol $[a,b]$ to denote the cycle in the class $[a,b]$ consisting of all unparameterized great circles on $S^{b+1} = \{x_0^2 + \cdots + x_{b+1}^2 = 1\}$ which meet the sphere $S^a = \{x_0^2 + \cdots + x_a^2 = 1\}$.

The cycles $[a, b]$ can be covered by chains $y_{a,b}$ in BS^n which project onto cycles $\bar{y}^*_{a,b}$ of $\bar{B}^*S^n$ and project further onto the cycles $[a, b]$ in such a way that every unparameterized great circle is covered by exactly one parameterized great circle with the possible exception of the boundary of the chain.

The chain $y_{a,b}$ is formed by all parameterized great circles on S^{b+1} which start on the half sphere $S^a \cap \{x_a \geqslant 0\}$ with initial direction tangent to the half sphere $S^{b+1} \cap \{x_{b+1} \geqslant 0\}$. Clearly $\bar{y}^*_{a,b}$ = image of $y_{a,b}$ under the quotient mapping $BS^n \to \bar{B}^*S^n$ is an $(a+b)$-cycle of $\bar{B}^*S^n$ and $\pi y^*_{a,b} = [a, b]$.

We agree to denote by $z_{a,b}$, $\bar{z}^*_{a,b}$ and $\{a, b\}$ the sets of circles lying over $y_{a,b}$, $\bar{y}^*_{a,b}$ and $[a, b]$ in the bundles α, $\bar{\alpha}^*$ and γ, respectively. That is, $\bar{z}^*_{a,b}$ and $\{a, b\}$ are the cycles of dimension $a + b + n - 1$ of $(\bar{\Lambda}^*S^n, \bar{\Lambda}^{*0}S^n)$ and $(\Gamma S^n, \Gamma^0 S^n)$, respectively, which are the images of $\bar{y}^*_{a,b}$ and $[a, b]$ under the Thom isomorphism.

4.10 Definition. Write $n = 2^k + s$, $0 \leqslant s < 2^k$, and put $2n - s - 1 = g(n)$. Consider the sequence of $g(n)$ cohomology classes of $\bar{B}^*S^n$ given by

$$(*)\quad u_1^{n-1} \cdot u_2^{n-1} = (u_1 + u_2)^{2n-2s-2} \cdot (u_1 \cdot u_2)^s, \ldots, (u_1 + u_2)^{2n-2s-2} \cdot (u_1 \cdot u_2),$$
$$(u_1 + u_2)^{2n-2s-2}, \ldots, (u_1 + u_2), 1.$$

By taking the cap product between a $(3n-3)$-cycle homologous to $\bar{z}^*_{n-1,n-1}$ and cocycles representing the elements of the sequence $(*)$, we obtain a sequence of $g(n)$ so-called subordinated cycles. Each of these cycles goes under $\pi : H_*(\bar{\Lambda}^*S^n, \bar{\Lambda}^{*0}S^n) \to (\Gamma S^n, \Gamma^0 S^n)$ into a cycle of $(\Gamma S^n, \Gamma^0 S^n)$. Note that every element in $(*)$ comes from an element in $H^*(\Delta S^n)$ under $\pi^* : H_*(\Delta S^n) \to H^*(\bar{B}^*S^n)$.

We abbreviate the sequence of such defined sets of $\bar{\Lambda}^*S^n$ by

$$(\dagger)\qquad \bar{\Lambda}^*_0, \ldots, \bar{\Lambda}^*_{2s-2}, \bar{\Lambda}^*_{2s}, \ldots, \bar{\Lambda}^*_{2n-3}, \bar{\Lambda}^*_{2n-2}.$$

Thus, the index l of $\bar{\Lambda}^*_l$ indicates that $\bar{\Lambda}^*_l$ is an $(l + n - 1)$-cycle of $(\bar{\Lambda}^*S^n, \bar{\Lambda}^{*0}S^n)$.

4.11 Definition. Let M be homeomorphic to S^n. This is equivalent to the existence of a homeomorphism $\phi : S^n \to M$ which is differentiable with the possible exception of one point. Then a differentiable homotopy equivalence $\phi : S^n \to M$ also exists (cf. Hirsch [**1**]). Such a ϕ determines a mapping $\Lambda\phi : \Lambda S^n \to \Lambda M$ which commutes with the S^1- and $\mathbf{Z}_2$-actions. Thus, in particular, $\Lambda\phi \mid \Lambda S^n$ induces mappings $(\bar{\Lambda}^*S^n, \bar{\Lambda}^{*0}S^n) \to (\bar{\Lambda}^*M, \bar{\Lambda}^{*0}M)$ and $(\Gamma S^n, \Gamma^0 S^n) \to (\Pi M, \Pi^0 M)$. Let Λ_l be the $\mathbf{Z}_2 \times \mathbf{Z}_2$-invariant covering in ΛS^n of the element $\bar{\Lambda}^*_l$ in 4.10(†). Put $\Lambda\phi \mid \Lambda_l = a_l$. The ϕ-family $\mathcal{Q}_l$ is now defined by $\{\phi_s a_l(\Lambda_l); s \geqslant 0\}$. The critical value of $\mathcal{Q}_l$ shall be denoted by κ_l.

4.12 Theorem. *Let M be homeomorphic to S^n. For the $g(n) = 2n - s - 1$ critical values κ_l of the ϕ-families $\mathcal{Q}_l$ as defined in* 4.11, *we have the relations* $0 < \kappa_0 \leqslant \cdots \leqslant \kappa_{2n-2}$. *If, for a subsequent pair $\{l, l'\}$, equality holds, i.e., $\kappa_l = \kappa_{l'}$ = (briefly) κ, then the number of critical S^1-orbits of E-value κ is infinite.*

Proof. $0 < \kappa_0$ follows as in the proof of 4.8—observe that $\phi_s a_0 : \Lambda_0 \to \Lambda M$ determines a homotopy equivalence $S^n \to M$. Note that the $(\mathbf{Z}_2 \times \mathbf{Z}_2)$-equivariant homology of $(\Lambda S^n, \Lambda^0 S^n)$ is the same as the homology of $(\bar{\Lambda}^*S^n, \bar{\Lambda}^{*0}S^n)$.

Assume $\kappa_l = \kappa_{l'} =$ (briefly) κ. We derive a contradiction from the assumption that there are only finitely many critical S^1-orbits of E-value κ. For that purpose we choose an open neighborhood $\mathfrak{U}$ of $\operatorname{Cr}\kappa$ by taking the union of small tubular neighborhoods of the critical S^1-orbits in $\operatorname{Cr}\kappa$. $\mathfrak{U}$ can be chosen invariant under the S^1- and $\mathbf{Z}_2$-actions; cf. the proof of 4.8. There exists an element $\phi_s a_{l'}(\mathrm{A}_{l'}) \subset \mathfrak{U} \cup \Lambda^{\kappa-}$. We also write $a_{l'}$. The counterimage of $a_{l'}$ of the open sets $\mathfrak{U}$ and $\Lambda^{\kappa-}$ in $\mathrm{A}_{l'}$ shall be denoted by $\mathcal{O}$ and $\mathfrak{N}$, respectively. $\{\mathcal{O}, \mathfrak{N}\}$ is an open covering of the $(\mathbf{Z}_2 \times \mathbf{Z}_2)$-invariant chain $\mathrm{A}_{l'}$ which is invariant under the $(\mathbf{Z}_2 \times \mathbf{Z}_2)$-action. Hence, $\{\bar{\mathcal{O}}^*, \bar{\mathfrak{N}}^*\}$ is an open covering of the cycle $\bar{\mathrm{A}}_l^*$.

Since $\mathfrak{U}$ contains no curve from some c to θc or $\theta' c$, $\bar{\mathcal{O}}^*$ contains no nontrivial cycle homologous to the cycle $\bar{y}_{0,1}^*$ with $\pi \bar{y}_{0,1}^* = [0, 1]$. That is to say, $(\pi^*(0,1) = u_1 + u_2) \,|\, \bar{\mathcal{O}}^* = 0$.

First consider the case $l' > 2s$, i.e., $\bar{\mathrm{A}}_l^* = \bar{\mathrm{A}}_{l'}^* \cap (u_1 + u_2)$. $\kappa_l = \kappa_{l'} = \kappa$ implies $(\pi^*(0,1) = u_1 + u_2) \,|\, \bar{\mathcal{O}}^* \neq 0$—a contradiction.

Assume $l' \leqslant 2s$, i.e., $\bar{\mathrm{A}}_l^* = \bar{\mathrm{A}}_{l'}^* \cap (u_1 \cdot u_2)$. $\kappa_l = \kappa_{l'} = \kappa$ implies $(\pi^*(1,1) = u_1 \cdot u_2) \,|\, \bar{\mathcal{O}}^* \neq 0$. That is to say, there must exist a 2-cycle $w_2 \in \pi\mathcal{O}$ on which $(1,1)$ does not vanish. To show that this is impossible we first observe that the $\mathbf{O}(2)$-bundle over w_2 must be orientable since $(0,1) \,|\, \pi\mathcal{O} = 0$. Denote its Euler class by e (cf. Steenrod [**1**]). We claim $e = 0$ and, hence, since $e \bmod 2 = (1,1)$, is the desired contradiction.

To prove the claim assume w_2 is connected. Then the S^1-bundle over w_2 is equivariantly mapped into a neighborhood $\mathfrak{U}(S^1.c)$ of a critical orbit $S^1.c$. If m is the multiplicity of c, $\mathfrak{U}(S^1.c)/\mathbf{Z}_m$ is a trivial S^1-bundle; the same, therefore, is also true for the S^1-bundle over $w_2/\mathbf{Z}_m$. Hence, $me = 0$. We need only show that e is infinitely cyclic. But this follows from the observation that w_2 is a cycle in $\tilde{\Delta} S^n$, i.e., a generator for $H_2(\tilde{\Delta} S^n)$, where $\tilde{\Delta} S^n$ is the 2-fold orientable covering of ΔS^n. $\square$

We conclude this section with an application of 4.12, just as we applied 4.8 to the proof of 4.9. The idea for this application is due to Thorbergsson [**1**] who considered $\frac{4}{9}$-pinched manifolds. His error in working with the space $(\Gamma S^n, \Gamma^0 S^n)$ instead of the space $(\bar{\mathrm{B}}^* S^n, \bar{\mathrm{B}}^{*0} S^n)$, as it must be, was corrected in Klingenberg [**4**]. That it actually suffices to consider a pinching of the form $\frac{1}{4} < K \leqslant 1$ was shown in Ballmann, Thorbergsson and Ziller [**1**].

The basis of the desired application is a refinement of the construction one makes in the usual proof of the sphere theorem: Let M be compact and simply connected with $\frac{1}{4} < K \leqslant 1$. Choose δ with $\frac{1}{4} < \delta < K \leqslant 1$. Let $\tilde{p} \in M$. Since the injectivity radius of M is $\geqslant \pi$, the distance sphere $\partial \bar{B}_{\pi/2\sqrt{\delta}}(\tilde{p})$ of radius $\pi/2\sqrt{\delta} < \pi$ around $\tilde{p}$ is a submanifold N. On N we have the antipodal mapping $p \to \bar{p}$. I.e., if $p = \exp_{\tilde{p}} X$ with $|X| = \pi/2\sqrt{\delta}$, $\bar{p} = \exp_{\tilde{p}}(-X)$.

4.13 Lemma. *Let M be compact, simply connected with $\frac{1}{4} < K \leqslant 1$. With the previously introduced notation, we consider, for a pair $(p, \bar{p})$ of antipodal points on N, the sets*

$$\bar{B}(p) = \{d(r,p) \leqslant d(r,\bar{p})\}; \quad \bar{B}(\bar{p}) = \{d(r,\bar{p}) \leqslant d(r,p)\}.$$

Then $\bar{B}(p) \subset B_\pi(p)$, $\bar{B}(\bar{p}) \subset B_\pi(\bar{p})$ and $E(p, \bar{p}) = E(\bar{p}, p) = \bar{B}(p) \cap \bar{B}(\bar{p})$ is a codimension one submanifold of M such that every geodesic of length $\pi/2\sqrt{\delta}$ starting from p or $\bar{p}$ meets $E(p, \bar{p})$ in exactly one point. This point is interior.

REMARK. For every $p \in N$, $E(p, \bar{p}) = E(\bar{p}, p)$ can thus play the role of an equator on M with $\bar{B}(p)$, $\bar{B}(\bar{p})$ as the corresponding hemispheres, just as in the usual proofs of the sphere theorem. That is to say, we show the existence of a full family of such equators and hemispheres for all pairs $(p, \bar{p})$ of the submanifold $N = \partial \bar{B}_{\pi/2\sqrt{\delta}}(\tilde{p})$ which may be viewed as the equator of the point $\tilde{p}$.

PROOF. Let $d(r, \bar{p}) \geqslant \pi/2\sqrt{\delta}$. We show that $d(r, p) < \pi/2\sqrt{\delta}$. Indeed, otherwise we would have a generalized triangle $rp\bar{p}$ where the sides c_{rp}, $c_{r\bar{p}}$ are minimizing geodesics and $c_{p\bar{p}}$ is the not necessarily minimizing geodesic of length $\pi/\sqrt{\delta}$ going from p through $\tilde{p}$ to $\bar{p}$. The length of this triangle is $\geqslant 2\pi/\sqrt{\delta} > 2\pi/\sqrt{K_0}$, $K_0 = \min K$, which contradicts Toponogov's triangle theorem. Thus, $\bar{B}(p) \subset B_{\pi/2\sqrt{\delta}}(p) \subset B_\pi(p)$ and, by the same argument, $\bar{B}(\bar{p}) \subset B_\pi(\bar{p})$.

Consider the differentiable functions

$$\Phi_p : r \in B_\pi(p) - \{p\} \mapsto d(p, r), \quad \Phi_{\bar{p}} : r \in B_\pi(\bar{p}) - \{\bar{p}\} \mapsto d(\bar{p}, r).$$

The unique minimizing unit speed geodesics from a point $r \in B_\pi(p) \cap B_\pi(\bar{p})$ to p and $\bar{p}$, respectively, have different initial directions. Thus, the function $\Phi = \Phi_{\bar{p}} - \Phi_p : \mathrm{B}_\pi(p) \cap B_\pi(\bar{p}) \to \mathbf{R}$ is regular. In particular, $\{\Phi = 0\} = E(p, \bar{p})$ is a submanifold of codimension 1. □

4.14 DEFINITION. Consider the unit sphere S^n in $\mathbf{R}^{n+1}$. A biangle b on S^n is a closed curve for which $b|[0, \frac{1}{2}]$ and $b|[\frac{1}{2}, 1]$ are half great circles. By $\tilde{\Lambda} S^n$ we denote the set of biangles, considered as a subspace of ΛS^n. $\tilde{\Lambda}_\theta S^n = \tilde{\Lambda} S^n \cap \Lambda_\theta S^n$ is the subspace of degenerate biangles b, i.e., $\theta b = b$ or $b(\frac{1}{4}) = b(\frac{3}{4})$. Let $\tilde{\alpha} : \tilde{\Lambda} S^n - \tilde{\Lambda}_\theta S^n \to \mathrm{B}S^n$ be the D^{n-1}-bundle where we associate to b the great circle c_0 with $c_0(0) = b(0)$, $c_0(\frac{1}{2}) = b(\frac{1}{2})$ and the line in $\mathbf{R}^{n+1}$ through $b(\frac{1}{4})$, $b(\frac{3}{4})$ parallel to the line through $c_0(\frac{1}{4})$, $c_0(\frac{3}{4})$. $\tilde{\alpha}$ commutes with the $\mathbf{Z}_2 \times \mathbf{Z}_2$-action, and thus we get from $\tilde{\alpha}$ the D^{n-1}-bundle $\bar{\tilde{\alpha}}^* : \bar{\tilde{\Lambda}}^* S^n - \bar{\tilde{\Lambda}}_\theta^* S^n \to \bar{\mathrm{B}}^* S^n$.

4.15 PROPOSITION. *$\tilde{\Lambda} S^n - \tilde{\Lambda}_\theta S^n$ is $(\mathbf{Z}_2 \times \mathbf{Z}_2)$-equivariantly homotopic to $\Lambda S^n - \Lambda^0 S^n$ such that the common subspace $\mathrm{B}S^n$ is invariant.*

This homotopy possesses a natural extension to a $(\mathbf{Z}_2 \times \mathbf{Z}_2)$-equivariant homotopy from $\tilde{\Lambda}_\theta S^n$ into $\Lambda_0 S^n$ inside the $(\mathbf{Z}_2 \times \mathbf{Z}_2)$-invariant set $\Lambda_\theta S^n$. Hence,

$$H^*\left(\bar{\tilde{\Lambda}}^* S^n, \bar{\tilde{\Lambda}}_\theta^* S^n\right) \cong H^*\left(\bar{\Lambda}^* S^n, \bar{\Lambda}^{*0} S^n\right).$$

In particular, we have in $\bar{\tilde{\Lambda}}^* S^n$ a sequence of $g(n)$ subordinated cycles $\bar{\tilde{\Lambda}}_l^*$, $l = 0, \ldots, 2n - 2$, which corresponds to the sequence $\bar{\Lambda}_l^*$, $l = 0, \ldots, 2n - 2$ (4.10(†)).

PROOF. To $b \in \tilde{\Lambda} S^n - \tilde{\Lambda}_\theta S^n$ associate the circle c with $c(\frac{1}{4}) = b(\frac{1}{4})$, $c(\frac{3}{4}) = b(\frac{3}{4})$ in the fibre $\alpha^{-1}(\tilde{\alpha}(b))$ over $\tilde{\alpha}(b)$. To $b \in \tilde{\Lambda}_\theta S^n$ associate the point circle $b(\frac{1}{4})$, $b(\frac{3}{4})$. Clearly, there exist $(\mathbf{Z}_2 \times \mathbf{Z}_2)$-deformations from b to its image c. □

We are now able to prove the application of 4.12 (due to Ballmann, Thorbergsson and Ziller [1]) which we mentioned earlier.

4.16 THEOREM. *Let M be compact, simply connected with $K_1/4 < K \leqslant K_1$, some $K_1 > 0$. Then there exist on M at least $g(n) = 2n - s - 1$ different simple closed geodesics with length in the interval $[2\pi/\sqrt{K_1}, 4\pi/\sqrt{K_1}[$.*

PROOF. We can assume $K_1 = 1$. Choose δ with $\frac{1}{4} < \delta < K \leqslant 1$. Our hypothesis implies the existence of a homeomorphism $\phi : S^n \to M$ with the following properties: If $\tilde{p}_1 = (0,\ldots,0,1)$ and $\tilde{q}_1 = (0,\ldots,0,-1) \in S^n$, ϕ is a diffeomorphism with the possible exception of $\tilde{q}_1$. The equator $N_1 = E(\tilde{p}_1, \tilde{q}_1)$ of the antipodal pair $(\tilde{p}_1, \tilde{q}_1)$ is mapped into the distance sphere $N = \partial\bar{B}_{\pi/2\sqrt{\delta}}(\tilde{p})$ of radius $\pi/2\sqrt{\delta}$ around $\tilde{p} = \phi(\tilde{p}_1)$. $\phi \mid N_1 : N_1 \to N$ commutes with the antipodal mappings on N_1 and N (cf. also 4.13).

This ϕ is constructed as in the sphere theorem; cf. also the proof of 4.9. More precisely, choose $\tilde{p} \in M$ and fix an isometry between $T_{\tilde{p}_1}S^n$ and $T_{\tilde{p}}M$. With the help of the exponential mapping, we define $\phi : \bar{B}_{\pi/2}(\tilde{p}_1) \to \bar{B}_{\pi/2\sqrt{\delta}}(\tilde{p})$. ϕ then commutes with the antipodal mapping on the boundary spheres $N_1 = \partial\bar{B}_{\pi/2}(\tilde{p}_1)$ and $N = \partial\bar{B}_{\pi/2\sqrt{\delta}}(\tilde{p})$.

Now let $\tilde{q} \in M$ have maximal distance from $\tilde{p}$. For $p \in N$, consider a triangle $\tilde{p}\tilde{q}p$. Since $\frac{1}{4} < \delta < K$, $d(\tilde{p}, \tilde{q}) < \pi\sqrt{\delta}$. With $d(\tilde{p}, p) = \pi/2\sqrt{\delta} < \pi$ we have $d(\tilde{q}, p) < \pi$; cf. the corresponding estimate in the proof of 4.13. Therefore, N belongs to the domain $B_\pi(\tilde{q})$ of normal coordinates based at $\tilde{q}$. We can thus extend ϕ to $\complement\,\bar{B}_{\pi/2\sqrt{\delta}}(\tilde{p})$ and get a homeomorphism from S^n to M with the desired properties, except that ϕ on the equator N_1 and in $\tilde{q}_1$ in general will not be differentiable; more precisely, the image of a half great circle from $\tilde{p}_1$ to $\tilde{q}_1$ on S^n is a once broken geodesic from $\tilde{p}$ to $\tilde{q}$ with the break at N. It is possible to smooth out ϕ at N_1. For our subsequent constructions, however, this is not really necessary.

In 4.15 we defined a sequence of $g(n)$ subordinated cycles in $(\bar{\tilde{\Lambda}}{}^*S^n, \bar{\tilde{\Lambda}}{}_0^*S^n)$ with the $(3n-3)$-dimensional cycle $\tilde{\Lambda}^*_{2n-2}$ formed by all biangles with initial point on the equator N_1 of S^n. We also use $\tilde{\Lambda}^*_{2n-2}$ to denote the $(\mathbf{Z}_2 \times \mathbf{Z}_2)$-invariant set in $\tilde{\Lambda}S^n$ covering this cycle. Our next goal is to construct a $(\mathbf{Z}_2 \times \mathbf{Z}_2)$-equivariant mapping $\tilde{a}_{2n-2} : \tilde{\Lambda}^*_{2n-2} \to \Lambda M$ having its image below the E-level $8\pi^2$. That is to say, the $g(n)$ critical $\mathbf{O}(2)$-orbits which we get according to 4.12 from the $g(n)$ subordinated cycles all yield closed geodesics of length $< 4\pi$. But such closed geodesics are simple, of length $\geqslant 2\pi$, and we thus get our theorem.

For the construction of $\tilde{a}_{2n-2}$ consider a pair $(p_1, \bar{p}_1)$ of antipodal points on the equator N_1 of S^n. Put $\phi(p_1) = p$. Then $\phi(\bar{p}_1) = \bar{p}$. Let $c_1(t)$, $0 \leqslant t \leqslant \pi$, be a half great circle from p_1 to $\bar{p}_1$. $c'(t) = \phi \circ c_1(t)$, $0 \leqslant t \leqslant \pi$, is a curve from p to $\bar{p}$. We define the geodesic segment c^* from p to the equator $E(p, \bar{p})$ as having initial direction $\dot{c}'(0)$ and going till the first intersection point r^* with $E(p, \bar{p})$ (cf. 4.13). Similarly, c^{**} shall be the geodesic starting from $\bar{p}$ with initial direction $-\dot{c}'(\pi)$ to the first intersection point r^{**} with the equator $E(p, \bar{p})$. Project the part of $c' - \{p\}$ which lies in $\bar{B}(p)$ from p into $E(p, \bar{p})$ and project the part of $c' - \{\bar{p}\}$ which lies in $\bar{B}(\bar{p})$ from $\bar{p}$ into $E(p, \bar{p})$. Adding the points r^*, r^{**} to the image, we get a curve $\tilde{c}(t)$, $0 \leqslant t \leqslant \pi$, on $E(p, \bar{p})$ from r^* to r^{**}. Put $\tilde{c}(\pi/2) = r$. Now deform the segment c^* from p to r^* into the segment from p to r by moving the endpoint $r^* = \tilde{c}(0)$ along $\tilde{c}(t)$, $0 \leqslant t \leqslant \pi/2$. Similarly, deform c^{**} into the segment from $\bar{p}$ to r by moving the endpoint $r^{**} = \tilde{c}(\pi)$ along $\tilde{c}(\pi - t)$, $0 \leqslant t \leqslant \pi/2$. Now define as the image of c_1 the once broken geodesic from p to $\bar{p}$ passing through r with parameter proportional to arc length. Since $E(p, \bar{p}) \subset B_\pi(p) \cap B_\pi(\bar{p})$, the length of the image curve is $< 2\pi$. Everything is done $(\mathbf{Z}_2 \times \mathbf{Z}_2)$-equivariantly and we thus get the desired mapping $\tilde{a}_{2n-2}$. □

Chapter 5. On the existence of infinitely many closed geodesics

In this chapter we give a survey on the results concerning the existence of infinitely many unparameterized closed geodesics. It would go far beyond the limitations of these notes to give complete proofs; we therefore restrict ourselves to a sketch of the basic ideas and refer the reader to the literature for details.

As stated in the introduction, the closed geodesics on a compact Riemannian manifold are in 1 : 1 correspondence to the periodic orbits of the geodesic flow $\phi_t : T_1M \to T_1M$ on the unit tangent bundle. ϕ_t is a special Hamiltonian flow. For Hamiltonian flows one knows that generically the periodic orbits are dense. This is a consequence of the Closing Lemma for Hamiltonian Systems (cf. Robinson [**1**]). However, no closing lemma is known for the subcategory of the geodesic systems. Nevertheless one may conjecture that generically also in this case the periodic orbits are dense. So far, this has been shown only for the geodesic flow on manifolds M with negative curvature. Clearly, if a compact differentiable manifold possesses a Riemannian metric g_0 with negative curvature, then all sufficiently nearly metrics g also have negative curvature. We thus have (cf. Anosov [**1**])

5.1 THEOREM. *Let M be a compact Riemannian manifold with strictly negative curvature K. Then the set* $\operatorname{Per} T_1M$ *of tangent vectors X with periodic orbit $\{\phi_t X; t \in \mathbf{R}\}$ is dense in T_1M.*

It follows that the underlying differentiable manifold possesses an open set of Riemannian metrics for which the above property is true.

PROOF. Anosov l.c. derives this from a general result on Hamiltonian systems of so-called hyperbolic type. A more elementary proof, which only employs the standard methods of Riemannian geometry, is contained in Klingenberg [**6**].

An essential step in this proof is the observation that the universal covering $\tilde{M}$ of M is of the type of a tangent space T_pM. $\tilde{M}$ possesses a canonical compactification $\operatorname{cp}\tilde{M}$ by points at infinity such that $(\operatorname{cp}\tilde{M}, \partial\operatorname{cp}\tilde{M} = \tilde{M}(\infty)) \cong (D^n, \partial D^n)$, $n = \dim M$. The fundamental group of M operates as a group Γ of isometries on $\tilde{M}$ which extends to $\operatorname{cp}\tilde{M}$. Every $\gamma \neq \mathrm{id}$ possesses, up to parameterization, exactly one invariant complete geodesic $\tilde{c}_\gamma$. The limit points $\alpha_\gamma = \lim_{t\to-\infty}\tilde{c}_\gamma(t)$ and $\omega_\gamma = \lim_{t\to\infty}\tilde{c}_\gamma(t)$ on $\tilde{M}(\infty)$ are the only invariant points for the infinite cycle group generated by γ. The theorem now essentially is a consequence of the fact that the set $\{(\alpha_\gamma, \omega_\gamma); \gamma \in \Gamma\}$ of pairs on $\tilde{M}(\infty)$ is dense in $\tilde{M}(\infty)$. For details see Klingenberg l.c. □

REMARK. The fundamental group $\pi_1 M$ of a compact manifold M of negative curvature is very large. As a measure for this largeness one introduces the so-called growth. Then one can show that $\pi_1 M$ has exponential growth (cf. Švarc [1] and Milnor [3]). On the other end of the spectrum are the compact manifolds M with finite fundamental group. Among these we mention in particular the compact manifolds with positive curvature.

As for the problem of the existence of an infinite number of prime closed geodesics on a compact manifold M, it does not matter whether one shows this for M or for a finite covering of M, since under the quotient of a finite group different prime closed geodesics cannot cover the same prime closed geodesic. For the remainder of this chapter, we therefore will restrict ourselves to simply connected manifolds.

The following result is of semilocal nature; we do not even need completeness of the manifold.

5.2 LEMMA. *Let c be a closed geodesic and assume that, for some $m \geqslant 1$,* index $c^m > 0$. *Then there exists $\alpha > 0$, $\beta \geqslant 0$ and $A > 0$ such that for all integers $m_0, m_1 \geqslant 1$,*

$$\operatorname{ind} c^{m_0} + m_1\alpha - \beta \leqslant \operatorname{ind} c^{m_0+m_1} \leqslant A(m_0 + m_1).$$

PROOF. This is a consequence of a result of Bott [1] who showed that index c^m is the sum of the ρ-index of c, $\rho^m = 1$; cf. Klingenberg [3] for a completion of Bott's result. One finds real numbers α_c, $\beta_c \geqslant 0$ such that

$$m\alpha_c - \beta_c \leqslant \operatorname{index} c^m \leqslant m\alpha_c + \beta_c \quad \text{for all } m \geqslant 1.$$

Now if index $c^m > 0$ for some m, $\alpha_c > 0$. With $\alpha = \alpha_c$, $\beta = 2\beta_c$, $A = \alpha_c + \beta_c$, one gets the desired estimates.

Next we write the Morse inequalities.

5.3 THEOREM. *Let $\kappa_0 < \kappa_1$ be regular values of $E: \Lambda M \to \mathbf{R}$. Assume that for each critical value $\kappa \in]\kappa_0, \kappa_1[$ the critical set* Cr κ *consists of a finite number of nondegenerate critical submanifolds. Then for the ith Betti numbers b_i we have*

$$b_i(\Lambda^{\kappa_1}, \Lambda^{\kappa_0}) \leqslant \sum_B b_i(N^-(B), \partial N^-(B)) = \sum_B b_{i-k(B)}(B).$$

Here, the sum Σ_B is taken over the finitely many nondegenerate critical submanifolds B with $E \mid B \in [\kappa_0, \kappa_1[$. $N^-(B)$ is the negative bundle over B and $k(B)$ is the index of B, i.e., the fibre dimension of $N^-(B)$. The Betti numbers are taken with respect to the $\mathbf{Z}_2$-homology.

REMARK. Actually, the Morse inequalities also hold with respect to homology with an arbitrary field of coefficients. Only in this case, the last equality may have to be replaced by $\leqslant$ if the negative bundle over B is not orientable.

PROOF. The basic idea of the proof is simple. Let $\kappa \in]\kappa_0, \kappa_1[$ be a critical value. Then κ is an isolated critical value, i.e., there exists $\varepsilon > 0$ such that κ is the only critical value in $[\kappa - \varepsilon, \kappa + \varepsilon]$. From the so-called Morse Lemma one has that $\Lambda^{\kappa+\varepsilon}$ is homotopically equivalent to $\Lambda^{\kappa-\varepsilon} \cup N^-(B_1) \cup \cdots \cup N^-(B_k)$ where the $B_1, \ldots, B_k$ are the nondegenerate critical submanifolds at the E-level κ. The symbol $\cup$ denotes an attaching of $\partial N^-(B_j)$ to $\Lambda^{\kappa-}$.

It follows that

$$H_*(\Lambda^{\kappa+\varepsilon}, \Lambda^{\kappa-\varepsilon}) = \sum_{j=i}^{k} H_*\big(N^-(B_j), \partial N^-(B_j)\big) = \sum_{j=i}^{k} H_{*-k(B_j)}(B_j)$$

The desired inequality for the Betti numbers now follows from the observation that, for two subsequent critical values $\kappa < \kappa'$, a cycle in $(N^-(B), \partial N^-(B))$, $E \mid B = \kappa$, may become a boundary by some cycle of $(N^-(B'), \partial N^-(B'))$, $E \mid B' = \kappa'$. Thus,

$$b_i(\Lambda^{\kappa'+\varepsilon'}, \Lambda^{\kappa-\varepsilon}) \leqslant b_i(\Lambda^{\kappa'+\varepsilon'}, \Lambda^{\kappa'-\varepsilon'}) + b_i(\Lambda^{\kappa+\varepsilon}, \Lambda^{\kappa-\varepsilon})$$

where $\kappa - \varepsilon < \kappa < \kappa + \varepsilon < \kappa' - \varepsilon' < \kappa' < \kappa' + \varepsilon'$ and κ, κ' are the only critical values in $[\kappa - \varepsilon, \kappa' + \varepsilon']$. For details see Seifert and Threlfall [**1**] and Milnor [**1**]. □

We now can prove the theorem of Gromoll and Meyer [**1**] for the nondegenerate case.

5.4 THEOREM. *Let M be a compact Riemannian manifold. Assume that the sequence $\{b_i\Lambda M\}$ of Betti numbers is unbounded. If the critical set* $\operatorname{Cr}\kappa$ *of $E\colon M \to \mathbf{R}$ can be decomposed into nondegeneratesummand critical submanifolds, then the number of prime critical orbits $S^1.c$ is infinite. That is to say, there exists on M an infinite number of unparameterized prime closed geodesics.*

PROOF. We derive a contradiction from the assumption that there exist only finitely many prime critical orbits $S^1.c_1, \ldots, S^1.c_l$. We first show: For every k, $1 \leqslant k \leqslant l$, there exists a constant R_k with the property

$$(*) \qquad \sum_{m=1}^{\infty} b_i\big(N^-(S^1.c_k^m), \partial N^-(S^1.c_k^m)\big) \leqslant R_k.$$

Indeed, observe first that in this sum b_i is $\neq 0$ only if $i = \operatorname{index} c_k^m =$ fibre dimension of $N^-(S^1.c_k^m)$ or $i = \operatorname{index} c_k^m - 1$, since the nondegenerate critical submanifold $S^1.c_k^m$ is a circle.

Assume now that for $m = m_0$ we have index $c_k^m = i$ or $= i - 1 > 0$. From 5.2 we have constants $\alpha = \alpha_k > 0$, $\beta = \beta_k \geqslant 0$ such that if $m_1 > (\beta + 2)/\alpha$, index $c_k^{m_0+m_1} > i + 1$. Thus, in $(*)$ there are at most $[(\beta + 2)/\alpha + 1]$ summands $\neq 0$, and each summand has value $\leqslant 1$. $[(\beta_k + 2)/\alpha_k + 1]$ then is the desired constant R_k.

Now let $i > \dim M$ such that $b_i\Lambda M > \sum_1^l R_k$. There exists a noncritical value $\kappa > 0$ such that index $c_k^m \leqslant i$ implies $E(c_k^m) < \kappa$. Hence,

$$\sum_1^l R_k < b_i\Lambda M = b_i\Lambda^\kappa M \leqslant \sum_{k=1}^{l} \sum_{m=1}^{\infty} b_i\big(N^-(S^1.c_k^m), \partial N^-(S^1.c_k^m)\big) \leqslant \sum_1^l R_k,$$

the desired contradiction. □

REMARKS. The Gromoll-Meyer hypothesis in 5.4 is satisfied whenever M is simply connected and has the homotopy type of a product $M' \times M''$ of two compact manifolds. Indeed, from Sullivan's [1] theory of the rational homotopy type follows: If M_0 is a compact, simply connected manifold then there exists an arithmetic sequence $\{2ar + b; r = 0, 1, \ldots\}$ such that $H_{2ar+b}(\Lambda M_0) \neq 0$. For instance, if M_0 is of the type of S^{n_0}, this sequence is given by $\{2(n_0 - 1)r + n_0 - 1\}$. If $\{2a'r + b'\}$ and $\{2a''r + b'\}$ are these arithmetic sequences for M' and M'', respectively, we consider

$$H_k(\Lambda(M' \times M'')) = H_k(\Lambda M' \times \Lambda M'') = \sum_{i=0}^{k} H_i(\Lambda M') \otimes H_{k-i}(\Lambda M'')$$

for $k = 2a'a''m + b' + b''$. On the right-hand side, the $m + 1$ summands $i = 2a'a''j + b'$ for $j = 0, \ldots, m$ are nonzero. Thus, $b_k \Lambda(M' \times M'') \geqslant m + 1$, i.e., the sequence $\{b_i \Lambda(M' \times M'')\}$ is unbounded.

On the other hand, if M is of the homotopy type of the sphere or one of the other simply connected symmetric spaces of rank 1, the sequence $\{b_i \Lambda M\}$ is bounded. Indeed, for $M = S^n$, the $\mathbf{Z}_2$-Betti numbers $b_i \Lambda S^n$ all are $\leqslant 2$, since the critical set on S^n decomposes into $\Lambda^0 S^n \cong S^n$ and $B_m = B_m S^n =$ space of m-fold covered great circles (cf. Chapters 3 and 4). index $B_m = (2m - 1)(n - 1)$.

$$H_*(B_m) = H_*(T_1 S^n) = H_*(S^n) \otimes H_*(S^{n-1}).$$

Hence, $H_i(N^-(B_m), \partial N^-(B_m)) = \mathbf{Z}_2$ for $i = (2m - 1)(n - 1) + j$, $j = 0, n - 1, n$ or $2n - 1$, and $= 0$ otherwise. From the Morse inequalities it follows that $b_i \Lambda S^n \leqslant 2$, all i.

We now consider a critical orbit $S^1.c$ which possibly is degenerate but isolated in Cr κ. To such an orbit one can again associate local homology groups. This is a consequence of a generalization of the Morse Lemma to the degenerate case.

To formulate this result let $T_c\Lambda = T_c^-\Lambda \oplus T_c^0\Lambda \oplus T_c^+\Lambda$ be the decomposition as in 3.3. Denote by $T_c'^0\Lambda \subset T_c^0\Lambda$ the subspace orthogonal to ∂c. Thus, index $c = \dim T_c^-\Lambda$, null $c = \dim T_c'^0\Lambda$. Let $\nu : N \to S^1$ be the normal bundle of the immersion $z \in S^1 \mapsto z.c \in M$. The above decomposition of $T_c\Lambda$ yields for ν the decomposition

$$\nu = \nu^- \oplus \nu^0 \oplus \nu^+ : N^- \oplus N^0 \oplus N^+ \to S^1.$$

On N we have the canonical S^1-action which respects the above decomposition.

Now consider a sufficiently small tubular S^1-invariant neighborhood U of $S^1.c$. Using the exponential mapping based on $S^1.c$, we can pull back the function $E : U \to \mathbf{R}$ to a function on a neighborhood D of the 0-section of N, which we again denote by E. Let D^-, D^0, D^+ be tubular neighborhoods of the zero section of the bundles ν^-, ν^0, ν^+ inside D.

5.5 LEMMA. *Let $S^1.c$ be an isolated critical orbit of index k and nullity l, $E \mid S^1.c = \kappa > 0$. Using the previously defined concepts, there exists a fibre preserving diffeomorphism $\psi : D \to D$ such that*

$$E \circ \psi(\xi) = \kappa - \|\xi^-\|_1^2 + \|\xi^+\|_1^2 + E_0(\xi^0),$$

where $\xi = (\xi^-, \xi^0, \xi^+) \in D^- \oplus D^0 \oplus D^+$ and $E_0(0) = DE_0(0) = D^2E_0(0) = 0$.

PROOF. If nullity $c = 0$, this is the equivariant Morse Lemma. It can be proved in the same way as the usual Morse Lemma in Hilbert manifolds (cf. Palais [**1**]). If nullity $c = l > 0$, one first constructs the so-called characteristic manifold $W_{ca} : D^0 \to D$ characterized by

$$W_{ca}(0_z) = z, \qquad T_{0_z}W_{ca}\left(T_{0_z}D^0\right) = T_zD^0,$$

and grad $E(W_{ca}(\xi^0))$ parallel to the subspace D^0. The functions

$$\eta \in D^- \oplus D^+ \mapsto F(\eta, \xi^0) = E\left(\eta + W_{ca}(\xi^0)\right) - E \circ W_{ca}(\xi^0)$$

have $\eta = 0$ as a nondegenerate critical point of index k for all sufficiently small $\xi^0 \in D^0$. A version of the proof of the equivariant Morse Lemma, dependent on the parameter ξ^0, yields the desired diffeomorphism ψ. □

From 5.5 we now get the local homology of an isolated critical orbit. With the above notation we write, with A in D or D^-, D^0, D^+, A^λ and $A^{\lambda -}$ for the sets $A \cap \{E \leqslant \lambda\}$ and $A \cap \{E < \lambda\}$, respectively. We denote by D_ε^-, D_ε^0, D_ε^+ the ε-neighborhoods in D^-, D^0, D^+, respectively.

5.6 LEMMA. *Let $S^1.c$ be an isolated critical orbit of E-value κ. With the previously introduced notation, there exists, for every sufficiently small $\varepsilon > 0$, a neighborhood $Z_\varepsilon \subset D_\varepsilon^- \oplus D_\varepsilon^0 \oplus D_\varepsilon^+$, where $Z_\varepsilon^{\kappa -} = (D_\varepsilon^- \oplus D_\varepsilon^0 \oplus D_\varepsilon^+)^{\kappa -}$, with the following property: $(Z_\varepsilon, Z_\varepsilon^{\kappa -})$ can be retracted S^1-equivariantly into $((D_\varepsilon^- \oplus D_\varepsilon^0)^\kappa, (D_\varepsilon^- \oplus D_\varepsilon^0)^{\kappa -})$ which in turn is homologically equivalent to the pair $((D_\varepsilon^0)^\kappa, (D_\varepsilon^0)^{\kappa -}) \times (D_\varepsilon^-, \partial D^-)$. Hence, if* index c = dim $D_\varepsilon^- = k$,

$$H_*(Z_\varepsilon, Z_\varepsilon^{\kappa -}) = H_{*-k}\left((D_\varepsilon^0)^\kappa, (D^0)^{\kappa -}\right)$$

only depends on the critical orbit $S^1.c$ and thus is an invariant of this orbit. We call

$$B_i(S^1.c) = \dim H_i(Z_\varepsilon, Z_\varepsilon^{\kappa -})$$

the ith type number of $S^1.c$ and

$$B_i^0(S^1.c) = \dim H_i\left((D_\varepsilon^0)^\kappa, (D_\varepsilon^0)^{\kappa -}\right)$$

the singular ith type number of $S^1.c$. Thus

$$B_i(S^1.c) = B_{i-k(c)}^0(S^1.c) \quad \text{if } k(c) = \text{index } c.$$

There are only finitely many i with $B_i^0(S^1.c) \neq 0$.

PROOF. See Klingenberg [**3**]. □

We also need the following result of Gromoll and Meyer [**1**].

5.7 LEMMA. *Let c be a prime closed geodesic and assume that for all $m = 1,2,\dots$ the critical orbits $S^1.c^m$ are isolated. Then there exist positive integers $\{m_1,\dots,m_s\}$, $s \leqslant 2^n$, $n = \dim M$, with $m_1 = 1$ such that, for any given m, there exists a well-determined $m_j \in \{m_1,\dots,m_s\}$ such that for all i, $B_i^0(S^1.c^m) = B_i^0(S^1.c^{m_j})$. It follows that there exists an i_0 and a constant $R = R(c)$ such that, for all $i \geqslant i_0$, $\sum_{m=1}^{\infty} B_i(S^1.c^m) \leqslant R$.*

PROOF. We first show: Let m_0 and $m_1 = qm_0$ be integers $\geqslant 1$ such that $\operatorname{null} c^{m_0} = \operatorname{null} c^{m_1}$. Then for all i, $B_i^0(S^1.c^{m_0}) = B_i^0(S^1.c^{m_1})$. To see this we replace the scalar product on $D(c^m)$ in the previous arguments by the scalar product

$$\langle \xi, \xi' \rangle_{m,1} = \langle \xi, \xi' \rangle_0 + m^{-2} \langle \nabla\xi, \nabla\xi' \rangle_0.$$

Clearly, these products are equivalent to the original one, and index and nullity of c^m do not change when we take the new scalar product. We define the characteristic manifolds and the type numbers with the new product and obtain the same results as before.

Now consider the mapping

$$(*) \qquad \xi \in D(c^{m_0}) \mapsto \xi^q \in D(c^{m_1}).$$

This is an isometry which carries $W_{ca}(c^{m_0})$ into $W_{ca}(c^{m_1})$. Thus, we get the equality of the singular type numbers. Now let $\rho = e^{2\pi i a}$, $\bar\rho = e^{-2\pi i a}$ be those eigenvalues of the Poincaré mapping P_ω associated to c which are roots of unity, i.e., $a = p/q$, p and q relatively prime. Denote by D the (possibly empty) set of denominators q of these a. There are at most n elements in D.

For each of the $s \leqslant 2^n$ subsets E of D let $m(E)$ be the least common multiple of the elements in E. For $E = \varnothing$ put $m(\varnothing) = 1$. We thus obtain a set $\{m_1 = 1,\dots,m_s\}$ of integers. For $m_j \in \{m_1,\dots,m_s\}$ we define N_j as the infinite set of the form $\{m_j q_{ji}, i = 1,2,\dots\}$ such that m_k does not divide $m_j q_{ji}$ for $k \neq j$.

These $N_1,\dots,N_s$ form a partition of the positive integers such that $m \in N_j$ implies $\operatorname{null} c^m = \operatorname{null} c^{m_j}$. This follows from the formula

$$\operatorname{null} c^m = \sum_{\rho^m = 1} \rho\text{-null } c.$$

Here, ρ-null c is the dimension of the space of complex valued Jacobi fields Y along c with $(Y(1), \nabla Y(1)) = \rho(Y(0), \nabla Y(0))$ (cf. Bott **[1]** and Klingenberg **[3]**).

Now let $i_0 > \dim M$ be so large that $B_i^0(S^1.c^{m_j}) = 0$ for all $i > i_0$, all $j = 1,\dots,s$. Either we have $k(c^m) = \operatorname{index} c^m = 0$ for all m—then $B_i(S^1.c^m) = B_i^0(S^1.c^m) = 0$ for all m and $i > i_0$ and we can choose $R(c) = 0$—or with $\alpha > 0$, $\beta \geqslant 0$ as in 5.2: if $B_i(S^1.c^{m_0}) = B^0_{i-k(c^{m_0})}(S^1.c^{m_0}) \neq 0$, and thus $i - k(c^{m_0}) < i_0$, then, for $m_1 \geqslant (\beta + i_0 + 1)/\alpha$, $k(c^{m_0+m_1}) > k(c^{m_0}) + i_0$ and, hence,

$$i - k(c^{m_0+m_1}) < i - i_0 - k(c^{m_0}) \leqslant 0.$$

Therefore, there are at most $l = [(\beta + i_0)/\alpha + 2]$ m's with $B_i(S^1.c^m) \neq 0$. If B is the maximum of the finitely many values of $B_i^0(S^1.c^m)$, $i = 0,\dots,i_0$, the number lB can serve as the desired upper bound $R(c)$.

We can now prove the Gromoll-Meyer theorem **[1]** in its general form.

5.8 THEOREM. *Assume that the sequence $\{b_i \Lambda M\}$ is unbounded. Then there exist on M infinitely many geometrically different closed geodesics.*

PROOF. We derive a contradiction from the assumption that there exist only finitely many prime critical orbits $\{S^1.c_1, \ldots, S^1.c_l\}$. Then for each $j = 1, \ldots, l$, the critical orbits $\{S^1.c_j^m,\ m = 1, 2, \ldots\}$ are isolated. From 5.7 we have constants $R(c_j)$ with

$$(*) \qquad \sum_{m=1}^{\infty} B_i\left(S^1.c_j^m\right) \leqslant R(c_j)$$

for all $i \geqslant$ some i_0. By hypothesis there exists an $i > i_0$, $i > \dim M$ with

$$(**) \qquad b_i \Lambda M > R(c_1) + \cdots + R(c_l).$$

On the other hand, $b_i \Lambda M$ is bounded from above by the sum of dimensions of the local i-dimensional homology at the various critical orbits $S^1.c_j^m$, i.e., by the sum of the $B_i(S^1.c_j^m)$, $j = 1, \ldots, l$, $m = 1, 2, \ldots$. Those $S^1.c_j^m$ with $B_i(S^1.c_j^m) \neq 0$ all lie below some noncritical E-value κ. Hence, using $(*)$,

$$b_i \Lambda M = b_i \Lambda^\kappa M = b_i(\Lambda^\kappa M, \Lambda^0 M) \leqslant R(c_1) + \cdots + R(c_l),$$

which contradicts $(**)$. □

As we already observed, the hypothesis in the Gromoll-Meyer theorem is not always satisfied. However, one can show that on a compact Riemannian manifold M with finite fundamental group there also exist infinitely many geometrically distinct closed geodesics if the sequence $\{b_i \Lambda M\}$ of Betti numbers is bounded. The critical step in the proof is the so-called Divisibility Lemma; see 5.20 below.

In these lectures we will restrict ourselves to the case where the underlying differentiable manifold of M is a sphere. We will also consider only the case that on M the critical set $\mathrm{Cr}(\Lambda M - \Lambda^0 M)$ is formed by nondegenerate critical S^1-orbits only.

We begin with the construction of certain rational cycles in ΩS^n.

5.9 DEFINITION. Let $\Omega S^n = (\Omega S^n, *)$ be the space of H^1-loops on $S^n = \{\Sigma_0^n x_i^2 = 1\}$ from $* = \{-1, 0, \ldots, 0\}$ to $*$. By $c_0(t)$ we denote the great circle on S^n in the (x_0, x_1)-plane with initial direction $X_0 = \dot{c}_0(0)/2\pi = (0, -1, \ldots, 0)$ in $T_* S^n$.

Identify $T_* S^n$ with the sphere $S'^{n-1} = \{\Sigma_1^n x_i^2 = 1\} \subset S^n$.

We consider the restriction of the unit tangent bundle $\tau_1 : T_1 S^n \to S^n$ to S'^{n-1}. The fibre $\tau_1^{-1}(X)$ over $X \in S'^{n-1} \equiv T_* S^n$ will be denoted by S_X^{n-1}.

(i) Define $v_1 : V_1 \equiv S_{X_0}^{n-1} \to \Omega S^n$ by associating to $\bar{p}$ the circle $c_{\bar{p}}$ with initial direction X_0 and $c_{\bar{p}}(\frac{1}{2}) = \bar{p}$. Note: If $\bar{p} = *$, $c_{\bar{p}}$ is the constant $=$ trivial circle $*$.

(ii) Define $v_2 : V_2 \equiv \tau_1^{-1}(S^{n-1}) \to \Omega S^n$ by associating to $\bar{p}'$ the circle $c_{\bar{p}'}$ having initial direction $X = \tau_1(\bar{p}') \in S^{n-1} \equiv T_* S^n$ with $c_{\bar{p}'}(\frac{1}{2}) = \bar{p}'$.

(iii) For each $r = 1, 2, \ldots$ define $v_{2r-1} : V_1 \times V_2^{r-1} \to \Omega S^n$ as the Ω-product $v_1 \cdot v_2 \cdot \ldots \cdot v_2$ ($(r-1)$ factors of v_2). By this we mean the composition of the circles $v_1(\bar{p}) \cdot v_2(\bar{p}^1) \cdot \ldots \cdot v_2(\bar{p}_{r-1}^1)$ with parameter proportional to arc length.

REMARK. The critical set of $E : \Omega S^n \to \mathbf{R}$ consists, besides the point $*$, of the critical manifolds $B_r^\Omega S^n$ of r-fold covered great circles starting and ending in $*$, $r = 1, 2, \ldots$. Put $E \mid B_r^\Omega S^n = 2\pi^2 r^2 = \kappa_r$. Then $B_r^\Omega S^n$ is nondegenerate of index $(2r-1)(n-1) = \text{index } c_0^r$.

5.10 PROPOSITION. *Under the inclusion* $i : (\Omega S^n, *) \to (\Lambda S^n, \Lambda^0 S^n)$, *the cycles* v_{2r-1} *are rationally nontrivial. At least for* $n \geqslant 4$, *they are homologous to the strong unstable manifolds* $W_{uu}(c_0^r) : T_{c_0^r}^- \Lambda \to \Lambda^{\kappa_r}$ *of* $c_0^r \in B_r S^n$.

Note. $W_{uu}(c_0^r)$ consists of the $c^* \in \Lambda S^n$ with $\lim \phi_s c^* = c_0^r$, for $s \to -\infty$. In particular, $T_{c_0^r} W_{uu}(c_0^r)$ coincides with $T_{c_0^r}^- \Lambda S^n$.

PROOF. We first show that $W_{uu}(c_0^r)$ has homology boundary zero in $(\Lambda S^n, \Lambda^0 S^n)$. Indeed the boundary, if it were nonzero, would have to be a relative cycle of dimension $(2r-1)(n-1)-1$. For $r = 1$ nothing has to be proved. We obtain an upper bound for the rational homology of B_r, $S^n \cong T_1 S^n$ by $H_*(S^n) \otimes H_*(S^{n-1})$. Thus, $H_i(B_r, S^n) \neq 0$ is possible only for $i = 0, n-1, n, 2n-1$. But for $n \geqslant 4$, $1 \leqslant r' < r$, the relation

$$(2r'-1)(n-1) + \{0, n-1, n, 2n-1\} = (2r-1)(n-1) - 1$$

is impossible.

Observe now that c_0^r is the only critical point in $i \circ v_{2r-1}$ at E-level κ_r. Thus, an application of the deformation ϕ_s, $s > 0$, will deform $i \circ v_{2r-1}$ into a cycle of $\{c_0^r\} \cup \Lambda^{\kappa_r -}$. But such a cycle is homologous to $W_{uu}(c_0^r)$.

It remains to show that the cycles $W_{uu}(c_0^r)$ are nonhomologous to zero. To see this simply observe that at no critical level $> \kappa_r$, there is a relative cycle having boundary dimension $= \dim W_{uu}(c_0^r)$. $\square$

We now introduce a homotopy of v_{2r-1}. Put $V_1 \times V_2^{r-1} = V_{2r-1}$.

5.11 DEFINITION. Define $h : [0, \pi] \times V_{2r-1} \to V_{2r+1}$ by $(\tau, \bar{p}, \bar{p}_1, \ldots, \bar{p}_{r-1}) \mapsto (*, \psi_\tau \bar{p}, \bar{p}_1, \ldots, \bar{p}_{r-1})$ with

$$\psi_\tau \bar{p} = \psi_\tau(x_0, 0, x_2, x_3, \ldots, x_n) = (x_0, \sin \tau x_1, \cos \tau x_2, x_3, \ldots, x_n).$$

That is, ψ_τ is the rotation by the positive angle τ in the (x_1, x_2)-plane.

With h we define

$$v'_{2r-1} = v_{2r+1} \circ h : V'_{2r-1} \equiv [0, \pi] \times V_{2r-1} \to \Omega S^n.$$

5.12 PROPOSITION.

(i) $$\partial v'_{2r-1} \approx -(\theta v_1) \cdot v_2^{r-1} - v_1 \cdot v_2^{r-1} \approx -\theta v_{2r-1} - v_{2r-1}.$$

(ii) *Using the homotopy from* $i \circ v_{2r-1}$ *into* $W_{uu}(c_0^r)$, *we get a homotopy*

$$W'_{uu}(c_0^r) : [0, \pi] \times T_{c_0^r}^- \Lambda \to \Lambda^{\kappa_r} S^n$$

with

$$W'_{uu}(c_0^r) \mid \{\tau\} \times T_{c_0^r}^- \Lambda = W_{uu}(\psi_\tau c_0^r), \qquad \partial W'_{uu}(c_0^r) \sim -\theta W_{uu}(c_0^r) - W_{uu}(c_0^r).$$

It follows that $\overline{W'_{uu}(c_0^r)}$ *(the* $^-$ *denotes the image in* $\bar{\Lambda} = \Lambda/_\theta \mathbf{Z}_2$*) is a* $\mathbf{Z}_2$*-cycle with* $\overline{W_{uu}(c_0^r)} = \overline{W'_{uu}(c_0^r)} \cap \omega$, *where* ω *is the canonical* 1*-cocycle in* $\bar{\Lambda} - \bar{\Lambda}_\theta$ *(cf.* 3.A.14).

PROOF. $\psi_\pi v_1(\bar{p})$ is the circle with initial direction $-X_0$ = initial direction of θc_0 passing through the point $\psi_\pi \bar{p} = (x_0, 0, -x_2, x_3, \ldots, x_n)$. Thus, $\psi_\pi v_1$ can be viewed as the composition of v_1 with the orientation reversing mapping $\psi_\pi : S^{n-1}_{X_0} \to S^{n-1}_{X_0}$ and θ. The homology of $\theta v_1 \cdot v_2^{r-1}$ with $\theta(v_1 \cdot v_2^{r-1})$ follows from the observation that the only critical point with E-value κ_r in the image is $\theta c_0^r = (\theta c_0)^r$. Apply ϕ_s, $s > 0$. Thus we have proved (i).

To prove (ii) we observe that the only critical point of E-value κ_r in the image of $v'_{2r-1} \mid \{\tau\} \times V_{2r-1}$ is $\psi_\tau c_0^r$, i.e., the r-fold covering of the great circle $\psi_\tau c_0$ with initial direction $(0, \cos\tau, \sin\tau, 0, \ldots, 0)$. □

For the next result we refer to the theory of dynamical systems with only hyperbolic singularities; cf. Klingenberg [3] and the references given there.

5.13 THEOREM. *Let $S^1.c$ be a nondegenerate critical orbit of $E : \Lambda M \to \mathbf{R}$ with E-value κ and* index $c = k$. *Then we have, for every $z.c \in S^1.c$, injective immersions*

$$W_{uu}(z.c) : T^-_{z.c}\Lambda \to \Lambda^\kappa M; \qquad 0^-_{z.c} \mapsto z.c,$$
$$W_{ss}(z.c) : T^+_{z.c}\Lambda \to \Lambda M - \Lambda^{\kappa -} M; \qquad 0^+_{z.c} \mapsto z.c,$$

with

$$T_{z.c} W_{uu}(z.c) = T^-_{z.c}\Lambda; \quad T_{z.c} W_{ss}(z.c) = T^+_{z.c}\Lambda.$$

These immersions commute with the S^1- and $\mathbf{Z}_2$-actions. Moreover,

$$c^* \in W_{uu}(c) \text{ iff } \lim_{s \to -\infty} \phi_s c^* = c, \qquad c^{**} \in W_{ss}(c) \text{ iff } \lim_{s \to +\infty} \phi_s c^{**} = c.$$

W_{uu} and W_{ss} are called strong unstable and strong stable manifolds. We also define the stable and unstable manifolds

$$W_u(S^1.c) : T^-(S^1.c) \to \Lambda^\kappa M, \quad W_s(S^1.c) : T^+(S^1.c) \to \Lambda M - \Lambda^{\kappa -} M$$

by

$$W_u(S^1.c) \mid T^-_{z.c}\Lambda = W_{uu}(z.c), \quad W_s(S^1.c) \mid T^+_{z.c}\Lambda = W_{ss}(z.c).$$

5.14 LEMMA. *Assume that all critical S^1-orbits in $\Lambda M - \Lambda^0 M$ are nondegenerate. Then there exists an arbitrarily small* $\mathbf{O}(2)$*-equivariant modification η^* of the vector field $\eta = -\operatorname{grad} E$ such that the stable manifolds W^*_s and strong unstable manifolds W^*_{uu}, defined as in* 5.13 *with η^* instead of η, have general equivariant intersection if* $\dim W^*_{uu} - \operatorname{codim} W^*_s \leqslant 2$.

REMARK. To explain the concept of general equivariant intersection, recall from the remark after 3.A.14 that for every positive integer m we have the embedding $m : \Lambda \to \Lambda_m \subset \Lambda$, where $mc = c^m$ is the m-fold covering of c. Λ_m consists of the elements with multiplicity divisible by m.

Now consider a $c^* \in W_{uu}(c') \cap W_s(S^1.c)$. Then the multiplicity $m(c^*)$ of c^* divides the multiplicities $m(c')$ and $m(c)$ of c' and c, respectively. Thus, $m(c^*)$ is a divisor of the greatest common divisor m of $m(c')$ and $m(c)$. Write $c' = c_0'^m$,

$c = c_0^m$, c_0' and c_0 not necessarily prime. For each divisor d of m, consider $W_{uu}(c_0'^d)$ and $W_s(S^1.c_0^d)$. Under $\bar{d} = m/d$, $W_{uu}(c_0'^d)$ is mapped into $W_{uu}(c)$ and $W_s(S^1.c_0^d)$ is mapped into $W_s(S^1.c)$, and the intersection of the images consists of the $c^* \in W_{uu}(c') \cap W_s(S^1.c)$ with $m(c^*)$ divisible by $\bar{d}$. That $W_{uu}(c')$ and $W_s(S^1.c)$ have general equivariant intersection now means that, for each divisor d of m, $W_{uu}(c_0'^d)$ and $W_s(S^1.c_0^d)$ have general equivariant intersection.

PROOF. Under our hypothesis, the critical values of E are isolated. Hence, if $W_{uu}(c') \cap W_s(S^1.c) \neq \varnothing$, there exist κ_0, κ_1 with $E(c) < \kappa_0 < \kappa_1 < E(c')$ such that $[\kappa_0, \kappa_1]$ contains no critical values. It is well known (cf. Milnor [**2**]) that then $\eta = -\text{grad}\, E$ possesses an arbitrarily small appropriate deformation η^* on a neighborhood of $\{\kappa_0 \leqslant E \leqslant \kappa_1\} \cap W_{uu}(c') \cap W_s(S^1.c)$ such that $W_{uu}^*(c')$ and $W_s^*(S^1.c)$, defined with η^*, have general intersection. That is, $W_{uu}^*(c') \cap W_s^*(S^1.c) = \varnothing$ if $\dim W_{uu}^*(c') - \operatorname{codim} W_s^*(S^1.c) = \operatorname{ind} c' - \operatorname{ind} c \leqslant 0$. If $\operatorname{ind} c' - \operatorname{ind} c = 1$, the intersection is either empty or consists of finitely many complete ϕ_s-orbits having c' and an element $z.c \in S^1.c$ as limit for $s \to \mp\infty$, respectively. If $\operatorname{ind} c' - \operatorname{ind} c = 2$ and the intersection is not empty, it consists of an open (i.e. without boundary points) 1-dimensional manifold of complete nonconstant ϕ_s-orbits having c' and a point on $S^1.c$ as limit for $s \to \mp\infty$, respectively. However, in our case we allow only $\mathbf{O}(2)$-equivariant modifications η^* of η, which poses certain restrictions, if the intersection contains elements with nontrivial isotropy group. As for the $\mathbf{Z}_2$-action, this causes no problem since on $\{\kappa_0 \leqslant E \leqslant \kappa_1\}$, $W_{uu}(c')$ and $W_s(S^1.c)$ are bounded away from the fixed point set of θ.

To handle the case of points with multiplicity > 1 on $W_{uu}(c') \cap W_s(S^1.c)$, denote by m the greatest common divisor of $m(c')$ and $m(c)$ and write $c' = c_0'^m$, $c = c_0^m$. The elements in $W_{uu}(c_0') \cap W_s(S^1.c_0)$ are all prime and, hence, η can equivariantly (= usually) be modified so as to bring these manifolds into general intersection. Extend the modification η^* of η to modifications $Td.\eta^*$ of $Td.\eta$ for all divisors d of m.

Now write $m = p_1 \cdot \ldots \cdot p_l$ as a product of primes. Let p be one of these primes. Then the set of elements in $W_{uu}^*(c_0'^p) \cap W_s^*(S^1.c_0^p)$ of multiplicity $\neq 1$ (in fact of multiplicity p) is given by $pW_{uu}^*(c_0') \cap pW_s^*(S^1.c_0)$, and here we have general intersection. Thus, general intersection has to be established only on the open set of points with multiplicity 1, but this clearly can be done equivariantly (= usually).

We now proceed by induction in assuming that we already know that $W_{uu}^*(c_0'^q)$ and $W_s^*(S^1.c_0^q)$ have general equivariant intersection for all q which are formed by the product of $\leqslant k$ of the primes $p_1, \ldots, p_l$. Let q' be the product of $k+1$ of these primes. Then, for every $c^* \in W_{uu}^*(c_0'^{q'}) \cap W_{ss}^*(c_0^{q'})$ with $m(c^*) > 1$, we have general equivariant intersection. Indeed, write $q' = rm(c^*)$. r is a product of $\leqslant k$ of the primes $p_1, \ldots, p_l$ and we already know that $W_{uu}(c_0'^r)$ and $W_s^*(S^1.c_0^r)$ have general equivariant intersection. Applying $m(c^*)$ to these manifolds, we get submanifolds of $W_{uu}^*(c_0'^{q'})$ and $W_s^*(S^1.c_0^{q'})$ of general equivariant intersection containing c^*. On the remaining open set, formed by the points of multiplicity 1, general equivariant intersection can also be achieved. □

REMARK. The property that all closed geodesics are nondegenerate is generic for the space of Riemannian metrics. If we require nondegeneracy only for the closed geodesics below a certain E-level, the property even is open and dense. For our subsequent applications, this would suffice. However, to simplify the formulation of our results, we do not adopt this approach.

5.15 DEFINITION. Assume that the critical S^1-orbits on $\Lambda M - \Lambda^0 M$ are nondegenerate and $-\text{grad } E$ has equivariantly been modified such that the unstable and stable manifolds have general equivariant intersection in the sense of 5.14. We denote these manifolds by W_u and W_s, respectively. ϕ_s denotes the flow with respect to the modified gradient vector field. An equivariant k-summit σ is defined as a k-dimensional subspace in the negative eigenspace $T_c^- = T_c^-\Lambda$ of a critical point c subject to the following conditions: σ is invariant under the isotropy group $I^\sim(c)$. Moreover, if $c = c_0^m$, c_0 prime, and $m = d\bar{d}$ with $\bar{d} > 1$, then σ contains the image $T\bar{d}\colon T_{c_0^d}^{\bar{d}} \to T_c^-$. In other words, σ is an affine subspace of T_c^- having general equivariant intersection with the origin of T_c^-. In particular, c is an equivariant summit when viewed as T_c^-. If $\sigma \subset c$, the multiplicity $m(\sigma)$ of σ is defined as the multiplicity $m(c)$ of c = order $I^\sim(c)$. For $k = 1, 2, \ldots$ denote by S_k the set of equivariant k-summits. For $k < 0$ put $S_k = \varnothing$. Let $\bar{S}_k$ be the free abelian group generated by S_k. In particular, $\bar{S}_k = 0$ for $k < 0$. For $k = 1, 2, \ldots$, denote by T_k the set of k-tunnels. Here, a k-tunnel τ is given by $I.\tilde{\sigma}$ where $\tilde{\sigma} \in S_{k-1}$ and $I = [z_0, z_1]$ is a nonconstant interval on S^1. We also allow $I = S^1$. For $k < 1$ put $T_k = \varnothing$. Let $\bar{T}_k$ be the free abelian group generated by T_k. In particular, $\bar{T}_k = 0$ for $k < 1$.

We now restrict ourselves to the orientable summits and tunnels, i.e., those summits and tunnels on which the isotropy group operates with positive determinant. Keeping the previously introduced notation, we choose orientations which commute with the S^1- and $\mathbf{Z}_2$-actions. As usual, $-\sigma$ denotes the element σ with orientation reversed. We can then define a boundary operator $\partial : \bar{S}_{k+1} + \bar{T}_{k+1} \to \bar{S}_k + \bar{T}_k$. Since on a tunnel $\tau = I.\sigma$, ∂ is operating as $\partial I.\sigma - I.\partial\sigma$, it suffices for $\sigma' \in S_{k+1}$ to define the integer coefficients a_σ and b_τ in

$$\partial\sigma' = \sum_\sigma a_\sigma \sigma + \sum_\tau b_\tau \tau; \qquad \sigma \in S_k, \tau \in T_k.$$

For that purpose observe that, if σ' is an equivariant subdisc of c', we have the restriction $W_{uu}(\sigma')$ of $W_{uu}(c')$. Consider a closed disc $\bar{D}_{uu}(\sigma')$ around (the origin of) σ'. $\dim(W_{uu}(\sigma') \cap W_s(S^1.c)) = 1$ means that $\bar{D}_{uu}(\sigma') \cap W_s(S^1.c)$ consists of finitely many points. For such a point p^* let $z.c$ be the limit of $\phi_s p^*$ for $s \to \infty$. Then $T\phi_s$, for $s \to \infty$, will transport the oriented tangent space $T_{p^*}\partial\bar{D}_{uu}(\sigma')$ into an oriented equivariant summit $\pm z.\sigma$ in $T_{z.c}^-$. $a_{z.\sigma} z.\sigma$ is now the sum of these $\pm z.\sigma$.

If $\dim(W_{uu}(\sigma') \cap W_s(S^1.\tilde{c})) = 2$, $\partial\bar{D}_{uu}(\sigma) \cap W_s(S^1.\tilde{c})$ is formed by finitely many open (i.e., without boundary points) connected curves k^*. Together with finitely many points p^* (or none at all), these k^* together form a closed curve S^* on $\partial\bar{D}_{uu}(\sigma')$. The limit $\phi_s p^*$, for $s \to \infty$, is an equivariant k-summit. The limit, for $s \to \infty$, of $\phi_s k^*$ has as closure an interval $I.\tilde{c} \subset S^1.\tilde{c}$. At the same time, $T\phi_s$

transports the oriented tangent space $T_{k*}\partial\overline{D}_{uu}(\sigma')$ into an equivariant oriented tunnel $\pm\tau$ in $T^-(I.\tilde{c})$. The sum of these $\pm\tau$ gives the term $b_\tau\tau$.

Such a differential complex is called the (oriented) Morse complex $\mathfrak{M} = \mathfrak{M}M$ of $\Lambda = \Lambda M$.

REMARKS. 1. The relation $\partial^2 = 0$ on $\mathfrak{M}$ follows from $\partial^2\overline{D}_{uu} = 0$. For a formal proof see below. There we also show that the Morse complex yields the homology of $(\Lambda M, \Lambda^0 M)$.

2. In $\partial\sigma'$ tunnels can occur. However, if $\tau = [z_0, z_1].\tilde{\sigma} \in \partial\sigma'$, and this is not $= S^1.\sigma'$, then $\partial\sigma'$ also contains summits which in their boundary contain the opposite of the summit boundary $-z_0.\tilde{\sigma} + z_1.\tilde{\sigma}$ of τ.

5.16 DEFINITION. Let the critical S^1-orbit in $\Lambda M - \Lambda^0 M$ be nondegenerate and consider the (oriented) Morse complex $\mathfrak{M} = \mathfrak{M}M$ as defined in 5.15. An (oriented, equivariant) k-manifold w in (Λ, Λ^0) is a subset of Λ which outside Λ^0 is an oriented submanifold of Λ of dimension k with singularities of codimension $\geqslant 2$. w shall be oriented and we permit the action of a finite subgroup of S^1 on w. The boundary ∂w of w shall be a $(k-1)$-manifold of (Λ, Λ^0) not necessarily connected.

Assume that w has general equivariant intersection with the stable manifolds W_s of codimension $\geqslant k-1$ and the same shall hold for ∂w with W_s of codimension $\geqslant k-2$. Then we define a mapping $K\colon (w, \partial w) \to (\overline{S}_k + \overline{T}_k, \overline{S}_{k-1} + \overline{T}_{k-1})$ in the Morse complex $\mathfrak{M}$. If $\dim(w \cap W_s(S^1.c)) = 0$, i.e., if $w \cap W_s(S^1.c)$ consists of points, consider for such a point p^* the limit $z.c \in S^1.c$ of $\phi_s p^*$ for $s \to \infty$. Then $T\phi_s$, for $s \to \infty$, will transport the oriented tangent space $T_{p*}w$ into an equivariant summit $\pm z.\sigma$ in $z.c$. The term $a_{z.\sigma}z.\sigma$ in Kw now is given as the sum of these $\pm z.\sigma$. If $\dim(w \cap W_s(S^1.c)) = 1$, $w \cap W_s(S^1.c)$ consists of finitely many open (i.e., without boundary points) connected curves k^*. Just as in 5.15 for the case $w = \partial\overline{D}_{uu}(\sigma')$, we associate to such a k^* an oriented equivariant tunnel $\pm\tau$ in $T^-(I.c)$. The term $b_\tau\tau$ in Kw is given by the sum of these $\pm\tau$. In the same manner we define $K\partial w$.

We finally extend the definition of K to k-manifolds w where only ∂w has general equivariant intersection with the W_s of codimension $\geqslant \dim\partial w - 1$. Before defining Kw, we bring w in general equivariant intersection with the W_s of codimension $\geqslant \dim w - 1$. If we do this in such a way as to keep ∂w fixed, then Kw does not depend on the manner in which this is done, provided the modification is sufficiently small.

REMARKS. 1. The verification of the last statement we leave to the reader.

2. The basic ideas for the concept of the Morse complex and the mapping K go back to Klingenberg [3]. In Klingenberg and Shikata [1] these ideas had been developed a little further, albeit under some restrictive assumptions. The present exposition owes much to the critical insight of N. Hingston.

3. In the concept of the Morse complex, we could also have allowed nonorientable equivariant summits and tunnels and then, in 5.16, nonorientable k-manifolds. But then everything has to be done with $\mathbf{Z}_2$-coefficients, of course.

4. The mappings $\phi_s : \Lambda \to \Lambda$ will also be called vertical, since the ϕ_s-flow lines are transversal to the sets $\{E = \text{const}\}$. We also call K vertical since K is closely related to the mapping ϕ_s for $s \to \infty$. On the other hand, the mappings $z : \Lambda \to \Lambda$, $z \in S^1$, and $\theta : \Lambda \to \Lambda$ will be called horizontal since they preserve E.

5.17 PROPOSITION. *Let the hypotheses of* 5.16 *hold and consider the mapping* K. *Let* $\bar{D}_{uu}(\sigma')$ *be a closed disc on* $W_{uu}(\sigma')$. *Then* $K\bar{D}_{uu}(\sigma') = \sigma'$, $K\partial\bar{D}_{uu}(\sigma') = \partial K\bar{D}_{uu}(\sigma')$. *Similarly, if* $\tau = I.\sigma$ *is an equivariant tunnel, let* $\bar{D}_u(\tau)$ *be* $= \{\bar{D}_{uu}(z.\sigma);\, z \in I\}$. *Then* $K\bar{D}_u(\tau) = \tau$ *and* $K\partial\tau = \partial K\tau$. *Hence,* $\partial^2 = 0$ *on* $\mathfrak{M}$. *Moreover,* $Kz = zK$, $z \in S^1$, *and* $K\theta = \theta K$.

PROOF. Write $\bar{D}$ instead of $\bar{D}_{uu}(\sigma')$. While $\partial\bar{D}$ has general equivariant intersection with the stable manifolds of codimension $\geqslant \dim\partial\bar{D} - 1$, we must apply to the interior of $\bar{D}$ a small 'lifting' to bring it in general equivariant intersection with the stable manifolds W_s of codimension $\geqslant \dim\bar{D} - 1$. The modified $\bar{D}$ will have nonempty intersection only with $W_s(S^1.c')$, and this at a single point p^* which, under ϕ_s, $s \to \infty$, goes into c'. Hence, $K\bar{D}_{uu}(\sigma') = \sigma'$. Similarly one sees that $K\bar{D}_u(\tau) = \tau$. $K\partial = \partial K$ is clear from the definitions and the same holds for $Kz = zK$, $K\theta = \theta K$. □

5.18 THEOREM. *Let the hypotheses of* 5.16 *hold. Denote by* Man $=$ Man(Λ, Λ^0) *the subcomplex of the singular complex* $S(\Lambda, \Lambda^0)$ *of* (Λ, Λ^0) *formed by the oriented k-manifolds as defined in* 5.16 *for* $k = 0, 1, \ldots$. Man$_k$ *denotes the subcomplex of* Man *formed by the j-manifolds with* $j \leqslant k$. $\mathfrak{M}_k$ *denotes the subcomplex of* $\mathfrak{M}$ *formed by the equivariant summits and tunnels of dimension* $\leqslant k$. *Then we have a mapping* $K :$ Man $=$ Man$(\Lambda, \Lambda^0) \to \mathfrak{M}$ *into the Morse complex* $\mathfrak{M}$ *with* im $K \mid$ Man$_k \subset \mathfrak{M}_k$. *$K$ commutes with ∂ and the z- and θ-actions,* $z \in S^1$. *The homology* H_*(Man) *of the complex* Man *coincides with the homology* $H_*(\Lambda, \Lambda^0)$ *of the singular complex* $S(\Lambda, \Lambda^0)$ *and K determines an isomorphism*

$$(*) \qquad K_* : H_*(\text{Man}) \cong H_*(\Lambda, \Lambda^0) \to H_*(\mathfrak{M}).$$

More precisely, for each $k = 0, 1, \ldots$ *we have a mapping* $J : \mathfrak{M}_k \to$ Man$_k$ *with* $K \circ J = \mathrm{id}_{\mathfrak{M}}$, $J \circ K = \mathrm{id}_{\text{Man}}$ *and* $(J\partial - \partial J)|_{\text{Man}_{k+1}} = 0 \bmod \text{Man}_{k-1}$. (*), *for* $H_* = H_{k+1}$, *then follows from the exact homology sequence of the triples* $(\mathfrak{M}_{k+1}, \mathfrak{M}_k, \mathfrak{M}_{k-1})$ *and* (Man$_{k+1}$, Man$_k$, Man$_{k-1}$).

PROOF. Let $\sigma' \in S_{k+1}$. Define $J\sigma'$ by $\bar{D}_{uu}(\sigma') =$ closed disc on $W_{uu}(\sigma')$ around $c' \supset \sigma'$. Similarly, define $J\tau$ by $\bar{D}_u(\tau)$ (cf. 5.17). Then clearly $K \circ J\sigma' = \sigma'$, $K \circ J\tau = \tau$ and $J \circ K\bar{D}_{uu}(\sigma') = \bar{D}_{uu}(\sigma')$, $J \circ K\bar{D}_u(\tau) = \bar{D}_u(\tau)$. From 5.17 we have $K\partial = \partial K$. Using the formula in 5.15 we have

$$J\partial\sigma' = \sum_{\sigma} a_\sigma \bar{D}_{uu}(\sigma) + \sum_{\tau} b_\tau \bar{D}_u(\tau),$$

while $\partial J\sigma' = \partial\bar{D}_{uu}(\sigma')$. Applying K to the right-hand side each time we get $\partial\sigma'$. Since $J \circ I.\sigma = I.J\sigma$, it follows that $K(J\partial(I.\sigma) - \partial J(I.\sigma)) = 0$ also. Hence, J induces a mapping $\mathcal{J}_* : H_{k+1}(\mathfrak{M}_{k+1}, \mathfrak{M}_k) \to H_{k+1}(\text{Man}_{k+1}, \text{Man}_k)$. We thus see that the exact homology sequences of the triples $(\mathfrak{M}_{k+1}, \mathfrak{M}_k, \mathfrak{M}_{k-1})$ and $(\mathrm{Man}_{k+1}, \mathrm{Man}_k, \mathrm{Man}_{k-1})$ are isomorphic in two of each three subsequent spots, the isomorphism being given by J_*, with K_* as its inverse. Therefore, also

$$H_{k+1}(\mathrm{Man}) = H_{k+1}(\mathrm{Man}_{k+1}, \mathrm{Man}_{k-1}) \cong H_{k+1}(\mathfrak{M}_{k+1}, \mathfrak{M}_{k-1}) = H_{k+1}(\mathfrak{M}).$$

But $H_{k+1}(\mathrm{Man}) = H_{k+1}(\Lambda, \Lambda^0)$ since every $(k+1)$-dimensional homology class in (Λ, Λ^0) can be represented by an element $w \in \mathrm{Man}_{k+1}$; cf. Sullivan **[2]** and also 2.18 and 3.15. □

5.19 PROPOSITION. *The cycles* $W_{uu}(c_0^r)$ *and their homotopies* $W'_{uu}(c_0^r)$ *under* $\pi: \Lambda M \to \Pi M$ *yield nontrivial* $\mathbf{Z}_2$*-cycles. Actually* $\pi W_{uu}(c_0^r) = \pi W'_{uu}(c_0^r) \cap \omega$, *where* ω *is the canonical* 1*-dimensional cocycle on* $\Pi S^n - \Pi_\theta^0 S^n$, *with* $\Pi_\theta S^n = \pi\Lambda_\theta S^n$, $\Lambda_\theta S^n$ *the fixed point set of* $\theta: \Lambda S^n \to \Lambda S^n$ (*cf.* 3.A.14).

Note. $W_{uu}(c_0^r)$ and $W'_{uu}(c_0^r)$ are invariant under $\mathbf{Z}_r \subset S^1$.

PROOF. We interpret the elements of ΠS^n as $\mathbf{O}(2)$-orbits in ΛS^n. We thus have to exclude a relation of the form $\mathbf{O}(2).c_0^r = \partial\mathbf{O}(2).y_1$, where y_1 is a 1-chain which we can assume to be transversal to $\mathbf{O}(2).c_0^r$. But such a relation clearly is impossible since the boundary of such a y_1 contains an even number of points.

REMARK. One can also show that $\pi W_{uu}(c_0^r)$ is an integer cycle of order 2. For a more detailed proof see Anosov **[2]**.

We now come to the main auxiliary result for our proof of the existence of an infinite number of unparameterized prime closed geodesics, the so-called Divisibility Lemma. In the proof we will consider simultaneously a submanifold w' with boundary $\partial w'$ and the images Kw' and $K\partial w'$ in the Morse complex $\mathfrak{M}$. While the statement of the result is about elements in $\mathfrak{M}$, in the proof the structure of w' will also be considered: The information we have on Kw', $K\partial w'$ will be used to make certain constructions on w', $\partial w'$.

5.20 DIVISIBILITY LEMMA. *Let* M *be a spherical manifold, i.e.,* M *is diffeomorphic to* S^n. *Assume that all closed geodesics on* M *are nondegenerate. Consider the Hilbert manifold* $\Lambda = \Lambda M$ *and the associated Morse complex* $\mathfrak{M} = \mathfrak{M}M$, *as defined in* 5.15. *For* $r = 1, 2, \ldots$ *we have in* (Λ, Λ^0) *the submanifolds* $W_{uu}(c_0^r)$ *and* $W'_{uu}(c_0^r)$ (*cf.* 5.11). *Bring these submanifolds into general equivariant position with respect to the stable manifolds of codimension* $\geqslant \dim W_{uu}(c_0^r) - 1$ *and* $\geqslant \dim W'_{uu}(c_0^r) - 1$ *and write* $w(r)$ *and* $w'(r)$ *instead. Thus,* $\partial w'(r) = -w(r) - \theta w(r)$. *Then there exists an equivariant summit* $\sigma'(r)$ *in* $Kw'(r)$ *and an equivariant summit* $\sigma(r)$ *in* $Kw(r)$, *contained with nonzero coefficients in* $\partial\sigma'(r)$, *such that the multiplicity* $m(\sigma'(r))$ *of* $\sigma'(r)$ *divides the multiplicity* $m(\sigma(r))$ *of* $\sigma(r)$, *briefly,* $m(\sigma'(r)) \mid m(\sigma(r))$.

Note. $\sigma'(r)$ $(\sigma(r))$ is contained in some $c'(r)$ $(c(r))$, and we may also write $m(c'(r)) \mid m(c(r))$.

PROOF. We fix an r and write $w', w, \sigma', \sigma, c', c, m, m'$ instead of $w'(r), w(r)$, $\sigma'(r)$, $\sigma(r)$, $c'(r)$, $c(r)$, $m(\sigma'(r)) = m(c'(r))$, $m(\sigma(r)) = m(c(r))$. Recall from 5.19 that there exists an equivariant summit σ in Kw such that $\tilde{\pi}\sigma$ has odd coefficient in $\tilde{\pi}Kw$ while $\tilde{\pi}\theta\sigma$ has even coefficient in $\tilde{\pi}Kw$. Here we have used $\tilde{\pi}$ to denote the mapping $\mathfrak{M} \to \mathfrak{M}/S^1$ of $\mathfrak{M}$ with respect to the S^1-action. If we write $\sigma \in \partial\sigma'$ we mean that σ occurs with nonzero coefficient in $\partial\sigma'$. Denote by $I^{\sim}(c')$ and $I^{\sim}(c)$ the isotropy group of c' (and also σ') and c (and also σ), respectively. Put $I^{\sim}(c')/I^{\sim}(c') \cap I^{\sim}(c) = I^{\sim}(c', c)$. Then $m' | m$ means that $I^{\sim}(c', c)$ is trivial. We will derive a contradiction from the assumption that, for all pairs $(\sigma, \sigma') \in (Kw, Kw')$ with $S^1.\sigma \cap \partial\sigma' \neq \varnothing$, we have $I^{\sim}(c', c) \neq \mathrm{id}$.

We first show—and this will turn out to be the decisive step in the proof—if there is a $\sigma' \in Kw'$ and a $\sigma \in \partial\sigma' \cap Kw$ such that $\partial\sigma' \cap Kw$ contains, besides σ, an element $z.\sigma \neq \sigma$, we get a contradiction. To see that, consider $Kw' | \sigma$ and $Kw' | z.\sigma$. We may assume $\sigma' \in Kw' | z.c$. The homotopy commutes with the S^1- and $\mathbf{Z}^2$-actions (see 5.11). This fact is preserved under vertical mappings. Hence, $\bar{z}.\sigma' = \sigma' \in \bar{z}.\ Kw' | z.c = Kw' | c$. Put $-\theta Kw \cap \partial Kw' | c = \sigma^*$. We now define a deformation of (Kw, Kw') as follows: Put $e^{2\pi it}.\sigma = \sigma(t)$ and $[1, e^{2\pi it}].\partial\sigma = \tau(t)$. Thus, $\sigma(t) - \tau(t)$ for $0 \leqslant t \leqslant t_0$, some small $t_0 > 0$, is a deformation of $\sigma = \sigma(0)$ obtained by dragging out σ along $[1, e^{2\pi it}]$ and thereby keeping the boundaries $\partial\sigma(t)$ and $\partial\sigma$ joined by the tunnel $\tau(t)$. Put $Kw - \sigma + \sigma(t) - \tau(t) = Kw(t)$. Thus, $Kw(t)$, $0 \leqslant t \leqslant t_0$, is a deformation of Kw. Similarly, define a deformation $Kw^*(t)$, $0 \leqslant t \leqslant t_0$, of $Kw^*(0) = -\theta Kw$ by

$$Kw^*(t) = -\theta Kw - \sigma^* + e^{2\pi it}.\sigma^* - [1, e^{2\pi it}].\partial\sigma^*.$$

We now use the same deformation to define a homotopy $Kw'(t)$ from $Kw(t)$ to $Kw^*(t)$, $0 \leqslant t \leqslant t_0$: On $Kw - \sigma$ let $Kw'(t)$ be given by Kw', whereas on $\sigma(t) - \tau(t)$, we define $Kw'(t)$ by $e^{2\pi it}Kw' | \sigma - [1, e^{2\pi it}]Kw' | \partial\sigma$. But now observe that, while σ' still is an element of $Kw'(t_0)$ (it occurs in $Kw' | z.\sigma$), the element $\sigma \in \partial\sigma'$ is missing—the desired contradiction. The possibility that σ is being cancelled by $-\sigma$ in the boundary of some newly created summit of $Kw'(t_0)$ can be discarded since such a summit would be of the form $e^{2\pi it_0}.\sigma_0'$ with $\sigma_0' \in Kw' | \sigma$.

Actually, the previously described modifications must be performed for (w', w) to give them a precise meaning. What is written above constitutes the 'program' in the Morse complex-image of what to do 'upstairs.' The implementation of this program requires only a number of rather routine steps which we will now describe. First, if $z.\sigma \subset z.c$, $\sigma' \subset c'$, we may assume we have a $q \in w \cap W_{ss}(z.c)$ and a $p' \in w' \cap W_{ss}(c')$ such that the homotopy $w' | q$ passes through p'. Similarly, with $\sigma \subset c$, $\sigma^* \subset c^*$, we have a $p \in w \cap W_{ss}(c)$ and a $p^* \in -\theta w \cap W_{ss}(c^*)$ such that the homotopy $w' | p$ joins p to p^*. Let $\overline{D}(p)$ be a small disc-like neighborhood of p on w with $K\overline{D}(p) = \sigma$.

The deformation $\sigma \mapsto \sigma(t) - \tau(t)$, $0 \leqslant t \leqslant t_0$, on Kw now amounts for w in the replacement of $\overline{D}(p)$ by $e^{2\pi it}\overline{D}(p)$ and at the same time joining $\partial\overline{D}(p)$ to $\partial e^{2\pi it}\overline{D}(p)$ by the tube $[1, e^{2\pi it}]\partial\overline{D}(p)$. We may assume that $w' | \overline{D}(p)$ is a homotopy from $\overline{D}(p)$ to a small disc-like neighborhood $\overline{D}(p^*)$ of p^* on $w^* = -\theta w$ with $K\overline{D}(p^*) = \sigma^*$. The deformation $Kw'(t)$, $0 \leqslant t \leqslant t_0$ of Kw' amounts to defining a homotopy of $e^{2\pi it}\overline{D}(p)$ by $e^{2\pi it}w' | \overline{D}(p)$ and defining a homotopy of the tube $[1, e^{2\pi it}]\partial\overline{D}(p)$ by $[1, e^{2\pi it}]w' | \partial\overline{D}(p)$.

Of course, now these deformations should be smoothed out and possible self-intersections should be removed in order to stay in the class Man. But clearly, there is enough room in the Hilbert manifold Λ to do so.

To complete the proof we first observe that in Kw' there must be at least one $\sigma' \in Kw'$ among the summits which have in their boundary an element on $S^1.\sigma$ for which $I^{\sim}(\sigma', \sigma) = I^{\sim}(c', c)$ has order > 2. Indeed, we know that $\tilde{\pi}\sigma$ has odd coefficient in $\tilde{\pi}Kw$, while its coefficient in $-\tilde{\pi}\theta Kw$ is even. The possibility that $S^1.\sigma \cap Kw'$ may contain tunnels does not invalidate the argument.

Thus consider $\sigma \in Kw$ such that $\tilde{\pi}Kw$ has odd coefficient at $\tilde{\pi}\sigma$. If σ does not occur as boundary element of some summit in Kw', then there must be a tunnel $\tau = -[1, z]\sigma$ or $= [z, 1].\sigma$ in Kw' such that $z.\sigma$ occurs in the boundary of some summit in Kw'. The summit part $-z.\sigma + \sigma$ of $\partial\tau$, together with this $z.\sigma$, then yields the element $\sigma \in Kw$. In such a case we modify (Kw', Kw) by removing from Kw' the tunnel τ and adding $\theta\tau$ to the right-hand boundary $-\theta Kw$ by identifying the element $\theta\sigma \in \partial\theta\tau$ with the element $-\theta\sigma$ in $-\theta Kw$. We call this the 'shift of the tunnel τ from the left to the right'. Clearly, $Kw' - \tau + \theta\tau$ can be viewed as a homotopy from $Kw - \partial\tau$ to $-\theta(Kw - \partial\tau)$.

We again write (Kw', Kw) for the modified chains and thus may assume that we have a summit σ' in Kw' with $\sigma \in \partial\sigma' \cap Kw$. From what we have shown above, σ is then the only element in $\partial\sigma' \cap Kw$ belonging to $S^1.\sigma$. We now 'shift σ' from the left to the right'. By this we mean that we remove σ' from Kw' and add $\theta\sigma'$ to the right-hand boundary $-\theta Kw$ of Kw' by identifying $\theta(\partial\sigma' \cap Kw)$ with $-\theta(\partial\sigma' \cap Kw) \in -\theta Kw$. Clearly, $Kw' - \sigma' + \theta\sigma'$ may be viewed as a homotopy from $Kw = \partial\sigma'$ to $-\theta(Kw - \partial\sigma')$. Now if $I^{\sim}(c', c)$ has order > 2 we are done: the right-hand boundary $-\theta(Kw - \partial\sigma')$ of the modified homotopy then contains $\geqslant 2$ elements of the $I^{\sim}(\theta c', \theta c) = I^{\sim}(c', c)$-orbit of a summit in $S^1.\theta\sigma \cap \partial\theta\sigma'$, which is impossible according to our previous arguments—just exchange the rôle of the left and right boundaries.

If $I^{\sim}(c', c)$ is of order 2, we continue with our constructions: First shift (if necessary) a tunnel from the left to the right and then a summit in the modified homotopy. In this way, we will eventually arrive at a summit σ' in the homotopy Kw' for which the isotropy group $I^{\sim}(c', c)$ has order > 2, and this yields the desired contradiction.

Here we have only given the 'program' in the Morse complex of what to do 'upstairs'. But it is quite obvious that the shift of a tunnel or a summit can be performed for w'—we therefore refrain from giving a detailed description of these operations. □

5.21 COMPLEMENT. *For a sequence $\{c(r), c'(r)\}$ of pairs of closed geodesics, containing the pairs $\{\sigma(r), \sigma'(r)\}$ from* 5.20, *we may assume $E(c(r)) < E(c'(r)) < E(c(r+1))$, $r = 1, 2, \ldots$.*

PROOF. The first inequality follows from $\sigma(r) \in \partial\sigma'(r)$. As for the second, observe that, according to Definitions 5.9 and 5.11, $w'(r)$ may be viewed as a restriction of $w(r+1)$. In the sequence $\{\sigma(r), \sigma'(r)\}$ choose the element $\sigma(r)$

such that it has maximal E-value among the summits in $Kw(r)$ with odd coefficient—there clearly exist such summits in $Kw(r)$. Then the coefficient of $\sigma'(r) \in Kw'(r)$ with $\sigma(r) \in \partial\sigma'(r)$ also must be odd. Such a $\sigma'(r)$ in $Kw'(r)$, when we want to obtain it by restricting $Kw(r+1)$, must come from taking the restriction on elements with odd coefficients in $Kw(r+1)$. □

We are now in a position to prove our main result.

5.22 THEOREM. *Let M be a spherical manifold with all closed geodesics nondegenerate. Then on M there exist infinitely many prime closed geodesics.*

REMARKS. 1. The conclusion also holds if the closed geodesics on M are degenerate. To derive a contradiction from the assumption that $\Lambda = \Lambda M$ carries only a finite number of prime critical orbits, all of their multiple coverings being isolated, one uses an approximation argument and Lemma 5.6 of Gromoll and Meyer [**1**] on the distribution of the type numbers in a tower of critical orbits. Actually, our methods yield the existence of infinitely many prime closed geodesics on every compact Riemannian manifold with finite fundamental group. For details see Klingenberg [**3**].

2. An entirely different proof that generically there exist infinitely many prime closed geodesics was given by N. Hingston [**1**].

PROOF. We derive a contradiction from the assumption that there exist on $\Lambda M - \Lambda^0 M$ only finitely many prime critical orbits, say $\{S^1.c_1, \ldots, S^1.c_l\}$. From 5.2 we then have $\alpha > 0$ and $\beta \geqslant 0$ such that

$$(*) \qquad \alpha(m' - m) - \beta \leqslant \operatorname{ind} c_j^{m'} - \operatorname{ind} c_j^{m}, \qquad j = 1, \ldots, l.$$

We also have an $A > 0$ (say $A = \max_j(\alpha_j + \beta_j)$) such that

$$(**) \qquad \operatorname{ind} c_j^m \leqslant Am, \qquad j = 1, \ldots, l.$$

Now consider the sequence $\{c(r), c'(r); r = 1, 2, \ldots\}$ of closed geodesics constructed in 5.20 and 5.21. For each fixed r_0 there exist $a = a(r_0)$, $b = b(r_0)$, $0 \leqslant a < b \leqslant l$, such that $c(r_0 + a)$ and $c(r_0 + b)$ have the same underlying prime closed geodesic. Hence, there exists a strictly increasing sequence $\{r(k), k = 1, 2, \ldots\}$, a fixed b, $0 < b \leqslant l$, and a prime c such that

$$c(r(k)) = c^{m(k)}, \qquad c(r(k) + b) = c^{\tilde{m}(k)}.$$

By choosing a subsequence of the sequence $\{r(k)\}$, which we again denote by $\{r(k)\}$, we also have a prime geodesic c' such that $c'(r(k)) = c'^{m'(k)}$. Put $E(c) = \kappa$, $E(c') = \kappa'$. From 5.20 we have $m(k) = q(k)m'(k)$ with integer $q(k)$. The relation

$$E(c(r(k))) < E(c'(r(k))) < E(c(r(k) + b))$$

(cf. 5.21) now reads

$$(***) \qquad m(k)^2\kappa < (m(k)/q(k))^2\kappa' < \tilde{m}(k)^2\kappa.$$

We claim that $\lim \tilde{m}(k)/m(k) = 1$ for $k \to \infty$. Indeed, from $(*)$ we have

$$\alpha(\tilde{m}(k) - m(k)) - \beta \leqslant \operatorname{ind} c^{\tilde{m}(k)} - \operatorname{ind} c^{m(k)}$$
$$= \dim w(r(k) + b) - \dim w(r(k)) = b(2n - 2) = \text{const.}$$

Moreover, $(**)$ shows that with $\operatorname{ind} c^{m(k)} = \dim w(r(k)) \to \infty$, for $k \to \infty$, $m(k) \to \infty$ also. From

$$1 < \tilde{m}(k)/m(k) \leqslant (\text{const} + \beta)/\alpha m(k) + 1,$$

we get our claim. Dividing $(***)$ by $m(k)$ we get, for $k \to \infty$,

$$\kappa \leqslant \left(\lim 1/q(k)^2\right)\kappa' \leqslant \kappa.$$

Hence, $q(k)$ assumes only finitely many integer values and we may assume $q(k) = q$, $\kappa = \kappa'/q^2$, which contradicts $(***)$. □

References

S. I. ALBER

[1] *On periodicity problems in the calculus of variations in the large*, Uspehi Mat. Nauk **12** (4) (1957), 57–125; English transl., Amer. Math. Soc. Transl. (2) **14** (1960), 107–172.

D. ANOSOV

[1] *Geodesic flows on closed riemannian manifolds with negative curvature*, Trudy Mat. Inst. Steklov **90** (1967); English transl., Proc. Steklov Inst. Mat., Amer. Math. Soc., Providence, R. I., 1969.

[2] *Some homology classes in the space of closed curves in the n-dimensional sphere*, Izv. Akad. Nauk SSSR Ser. Mat. **45** (1981) = Math. USSR Izv. **18** (1982), 403–422.

W. BALLMANN, G. THORBERGSSON AND W. ZILLER

[1] *Closed geodesics on positively curved manifolds*, Bonn and Philadelphia, 1981, preprint; to appear in J. Differential Geom.

A. BOREL

[1] *La cohomologie* mod 2 *des certaines espaces homogènes*, Comment. Math. Helv. **27** (1953), 165–197.

R. BOTT

[1] *On the iteration of closed geodesics and the Sturm intersection theory*, Comm. Pure Appl. Math. **9** (1956), 171–206.

S. S. CHERN

[1] *On the multiplication of the characteristic ring of a sphere bundle*, Ann. of Math. (2) **49** (1948), 362–372.

J. DIEUDONNÉ

[1] *Foundations of modern analysis*, Academic Press, New York and London, 1960.

H. ELIASSON

[1] *On the geometry of manifolds of maps*, J. Differential Geom. **1** (1967), 165–194.

P. FLASCHEL UND W. KLINGENBERG

[1] *Riemannsche Hilbertmannigfaltigkeiten. Periodische Geodätische*, Lecture Notes in Math., 282, Springer, Berlin and New York, 1972.

D. GROMOLL AND W. MEYER

[1] *Periodic geodesics on compact Riemannian manifolds*, J. Differential Geom. **3** (1969), 493–510.

K. GROVE

[1] *Condition* (C) *for the energy integral on certain path spaces and applications to the theory of geodesics*, J. Differential Geom. **8** (1973), 207–223.

N. HINGSTON

[1] *Equivariant Morse theory and closed geodesics*, Preprint, Dept. of Math., Univ. of Pennsylvania, Philadelphia, Pa., 1982.

M. HIRSCH

[1] *Differential topology*, Springer, New York and Berlin, 1976.

W. KLINGENBERG

[1] *Simple closed geodesics on pinched spheres*, J. Differential Geom. **2** (1968), 225–232.

[2] *Der Indexsatz für geschlossene Geodätische*, Math. Z. **139** (1974), 231–256.

[3] *Lectures on closed geodesics*, Grundlehren Math. Wiss., Bd. 230, Springer, Berlin and New York, 1978.

[4] *On the existence of closed geodesics on spherical manifolds*, Math. Z. **176** (1981), 319–325.

[5] *Über die Existenz unendlich vieler geschlossener Geodätischer*, Akad. Wiss. Lit. Mainz. Abh. Math.-Naturwiss. Kl. No. 1, 1981.

[6] *Riemannian geometry*, Studies in Math., No. 1, de Gruyter, Berlin and New York, 1982.

W. KLINGENBERG AND T. SAKAI

[1] *Injectivity radius for* $\frac{1}{4}$*-pinched manifolds*, Arch. Math. **34** (1980), 371–376.

W. KLINGENBERG AND Y. SHIKATA

[1] *On the existence theorem of infinitely many closed geodesics* (Proc. Internat. Conf. on Topology, Moscow 1979), Trudy Mat. Inst. Steklov, No. 159, 1981.

M. KLINGMANN

[1] *Das Morse'sche Indextheorem bei allgemeinen Randbedingungen*, J. Differential Geom. **1** (1967), 371–380.

L. LYUSTERNIK

[1] *The topology of function spaces and the calculus of variations in the large*, Trudy Mat. Inst. Steklov **19** (1947); English transl., *The topology of the calculus of variations in the large*, Transl. Math. Mono., Vol. 16, Amer. Math. Soc., Providence, R. I., 1966.

L. LYUSTERNIK AND A. I. FET

[1] *Variational problems on closed manifolds*, Dokl. Akad. Nauk SSSR **81** (1951), 17–18.

J. MILNOR

[1] *Morse theory*, Ann. Math. Studies, no. 51, Princeton Univ. Press, Princeton, N. J., 1963.

[2] *Lectures on the h-cobordism theorem*, Princeton Math. Notes, Princeton Univ. Press, Princeton, N. J., 1965.

[3] *A note on curvature of the fundamental group*, J. Differential Geom. **2** (1968), 1–7.

M. MORSE

[1] *Calculus of variations in the large*, Amer. Math. Soc. Colloq. Publ., vol. 18, Amer. Math. Soc., Providence, R. I., 1934.

[2] *Generalized concavity theorems*, Proc. Nat. Acad. Sci. U.S.A. **21** (1935), 353–362.

R. PALAIS

[1] *Morse theory on Hilbert manifolds*, Topology **2** (1963), 299–340.

R. PALAIS AND S. SMALE

[1] *A generalized Morse theory*, Bull. Amer. Math. Soc. **70** (1964), 165–172.

C. ROBINSON

[1] *Generic properties of conservative systems*, Amer. J. Math. **92** (1970), 562–603.

H. SEIFERT AND W. THREFALL

[1] *Variationsrechnung im Grossen*, Teubner, Leipzig, 1938.

J.-P. SERRE

[1] *Homologie singulière des espaces fibrés. Applications*, Ann. of Math. (2) **54** (1951), 425–505.

E. SPANIER

[1] *Algebraic topology*, McGraw-Hill, New York and London, 1966.

J. SOLÁ-MORALES

[1] *Onthe continuation of the* –grad *E flow of* $H^1(S^1, M)$, Arch. Math. (Basel) **34** (1980), 140–142.

N. STEENROD

[1] *The topology of fibre bundles*, Princeton Univ. Press, Princeton, N. J., 1951.

D. SULLIVAN

[1] *Differential forms and the topology of manifolds*, Manifolds-Tokyo 1974 (A. Hattori, ed.), Univ. of Tokyo Press, Tokyo, 1975.

[2] *Singularities in spaces*, Lecture Notes in Math., 209, Springer, Berlin and New York, 1971, pp. 196–206.

A. S. ŠVARC

[1] *A volume invariant of coverings*, Dokl. Akad. Nauk SSSR **105** (1955), 32–34. (Russian)

G. THORBERGSSON

[1] *Non-hyperbolic closed geodesics*, Math. Scand **44** (1979), 135–148.

ABCDEFGHIJ–AMS–89876543

[1] [illegible] of the fundamental group [illegible]

N. Mori

[1] [illegible] Math. Soc. [illegible]

[2] [illegible] 35 [illegible]

[illegible]

K. P[illegible]

[1] [illegible] Sci. [illegible]

[illegible]

[1] [illegible] 2 [illegible]

[illegible]

[1] [illegible]

[illegible]

[1] [illegible] (2) [illegible]

[illegible]

[illegible]

[1] [illegible]

[illegible]

[1] [illegible]

[illegible]

[1] [illegible] 1954 [illegible]

I. Schur [illegible]

[1] [illegible] 1974 [illegible]

Edition, [illegible]

[2] [illegible]

[illegible] Zeit. [illegible]

A. S. Svarc

[1] A volume invariant of [illegible]

(Russian)

[illegible]

[1] [illegible]